D0944832

Surface Infrared and Raman Spectroscopy

Methods and Applications

METHODS OF SURFACE CHARACTERIZATION

Series Editors:

Cedric J. Powell, *National Institute of Standards and Technology, Gaithersburg, Maryland*
Alvin W. Czanderna, *Solar Energy Research Institute, Golden, Colorado*
David M. Hercules, *University of Pittsburgh, Pittsburgh, Pennsylvania*
Theodore E. Madey, *Rutgers, The State University of New Jersey,*
New Brunswick, New Jersey
John T. Yates, Jr., *University of Pittsburgh, Pittsburgh, Pennsylvania*

Volume 1   VIBRATIONAL SPECTROSCOPIES
OF MOLECULES ON SURFACES
Edited by John T. Yates, Jr., and Theodore E. Madey

Volume 2   ION SPECTROSCOPIES FOR SURFACE ANALYSIS
Edited by A. W. Czanderna and David M. Hercules

Volume 3   SURFACE INFRARED AND RAMAN SPECTROSCOPY
Methods and Applications
W. Suëtaka with the assistance of John T. Yates, Jr.

A Continuation Order Plan is available for this series. A continuation order will bring delivery of each new volume immediately upon publication. Volumes are billed only upon actual shipment. For further information please contact the publisher.

Surface Infrared and Raman Spectroscopy

Methods and Applications

W. Suëtaka

Professor Emeritus
Tohoku University
Tsuchiura, Japan

With the assistance of

John T. Yates, Jr.

University of Pittsburgh
Surface Science Center
Pittsburgh, Pennsylvania

PLENUM PRESS • NEW YORK AND LONDON

Library of Congress Cataloging-in-Publication Data

On file

ISBN 0-306-44963-3

© 1995 Plenum Press, New York
A Division of Plenum Publishing Corporation
233 Spring Street, New York, N. Y. 10013

10 9 8 7 6 5 4 3 2 1

Printed in the United States of America

About the Series

A large variety of techniques are now being used to characterize many different surface properties. While many of these techniques are relatively simple in concept, their successful utilization involves employing rather complex instrumentation, avoiding many problems, discerning artifacts, and carefully analyzing the data. Different methods are required for handling, preparing, and processing different types of specimen materials. Many scientists develop surface characterization methods, and there are extensive developments in techniques reported each year.

We have designed this series to assist newcomers to the field of surface characterization, although we hope that the series will also be of value to more experienced workers. The approach is pedagogical or tutorial. Our main objective is to describe the principles, techniques, and methods that are considered important for surface characterization, with emphasis on how important surface characterization measurements are made and how to ensure that measurements and interpretations are satisfactory, to the greatest extent possible. At this time, we have planned four volumes, but others may follow.

The first volume brought together a description of methods for vibrational spectroscopy of molecules on surfaces. Most of these techniques are still under active development; commercial instrumentation is not yet available for some techniques, but this situation could change in the next few years. The current state of the art of each technique was described as were their relative capabilities. An important component of the first volume was the summary of the relevant theory.

This book is the first of two volumes that contain descriptions of the techniques and methods of electron and ion spectroscopies which are in widespread use for surface analysis. These two volumes are and will be largely concerned with techniques for which commercial instrumentation is available. The books are intended to fill the gap between a manufacturer's handbook, and review articles that highlight the latest scientific developments.

A fourth volume will deal with techniques for specimen handling, beam artifacts, and depth profiling. It will provide a compilation of methods that have proven useful for specimen handling and treatment, and it will also address the common artifacts and problems associated with the bombardment of solid surfaces by photons, electrons, and ions. A description will be given of methods for depth profiling.

Surface characterization measurements are being used increasingly in diverse areas of science and technology. We hope that this series will be useful in ensuring that these measurements can be made as efficiently and reliably as possible. Comments on the series are welcomed, as are suggestions for volumes on additional topics.

C. J. Powell
Gaithersburg, Maryland
A. W. Czanderna
Golden, Colorado
D. M. Hercules
Pittsburgh, Pennsylvania
T. E. Madey
New Brunswick, New Jersey
J. T. Yates, Jr.
Pittsburgh, Pennsylvania

Preface

Infrared (IR) and Raman spectroscopy may seem to be old-fashioned approaches to surface characterization because of the emergence of many modern techniques to study surfaces. However, these techniques provide detailed molecular-level information about the surface species present in ultrahigh vacuum environments as well as in various absorbing media. Furthermore, many variations have been developed in recent years and are still emerging. As a result, IR and Raman spectroscopy have been greatly improved in sensitivity and today are easily applied to the *in situ* and real-time observation of solid surfaces in liquid and gaseous media as well as of solid/solid interfaces. For this reason, these techniques are widely employed in the investigation of heterogeneous catalysis, electrode processes, growth of thin films, corrosion and its inhibition, lubrication, adhesion, and surface treatment of engineering materials.

There are many excellent books that treat surface IR or Raman spectroscopy from the viewpoint of pure science. However, there exist only a few books that treat both IR and Raman spectroscopies for engineers and researchers in applied science using the best knowledge of the author. This book is written, therefore, for those who are not spectroscopists but who are working or beginning to work with IR and Raman spectroscopy in the investigation of surfaces of practical materials. Consequently, this book does not contain detailed theoretical treatments, and only basic knowledge of vibrational spectroscopy is required for understanding the discussions in this book. Applications of IR and Raman spectroscopy to the investigation of a variety of real surfaces are illustrated. Typical results obtained for species on clean and well-defined surfaces in ultrahigh vacuum environments are also included in this book for better understanding of the basic principles governing various surface phenomena.

I should like to express my most heartfelt thanks to Professor N. Sheppard for encouraging me to write this book and for kind arrangements for publication. I am particularly indebted to Professor J. T. Yates, Jr., who read the entire manuscript and pointed out a number of mistakes. His detailed and perspicacious comments and suggestions have substantially improved the manuscript. The editorial staff of Plenum Publishing Corporation provided valuable advice all along the route from manuscript to printed volume.

I am also most grateful to the following scientists for supplying the reprints and preprints: Professor A. M. Bradshaw (Berlin), A. Campion (Austin), Dr. Y. J. Chabal (Murray Hill), Professor J.-N. Chazalviel (Palaiseau), Dr. M. K. Debe (St. Paul), Dr. J. G. Gordon II (San Jose), Professor R. G. Greenler (Milwaukee), Professor A. Hatta (Sendai), Dr. F. M. Hoffman (Annandale), Professor H. Ishida (Cleveland), Professor C. Lamy (Poitiers), Dr. C. A. Melendres (Argonne), Professor H. Metiu (Santa Barbara), Professor M. Osawa (Sendai), Professor A. Otto (Düsseldorf), Professor N. Sheppard (Norwich), Dr. J. D. Swalen (San Jose), Professor R. G. Tobin (East Lansing), Professor Vo-Van Truong (Moncton) and Professor J. T. Yates, Jr. (Pittsburgh). Thanks are also due to the following publishers for permission to reproduce figures and tables appearing in this book: Academic Press Inc., American Chemical Society, American Institute of Physics, Chemical Society of Japan, Éditions Gauthier-Villars, Elsevier Sequois S. A., Elsevier Science Publishers B.V., Industrial Publishing & Consulting, Inc., Japanese Society of Tribologists, Japan Institute of Metals, Japan Society of Analytical Chemistry, Pergamon Press Ltd., Plenum Publishing Corporation, Publication Office of Japanese Journal of Applied Physics, The Royal Society, Society for Applied Spectroscopy and Spectroscopical Society of Japan.

Tsuchiura, Japan

Prefatory Note

The surface vibrational spectroscopies provide incisive information about the chemical species present on surfaces through the use of group frequency assignment methods. This book deals with many of the optical vibrational spectroscopy methods which are employed by research workers and technicians in the field. This book is a fitting sequel to *Vibrational Spectroscopy of Molecules on Surfaces* in this series, Methods of Surface Characterization, Volume 1, Plenum Press, New York (1987), edited by J. T. Yates, Jr. and T. E. Madey. The reader will find excellent descriptions of the physical basis for the various optical methods as well a many examples of the use of these methods to solve problems of importance in a variety of technologies.

John T. Yates, Jr.

Pittsburgh, Pennsylvania

Contents

1

Introduction

1. Surface Analysis

The knowledge of the submicroscopic structure of surfaces, adsorbed layers, deposited surface films, and interfaces is of importance for the fundamental understanding of physical, mechanical, and chemical properties of solid materials. At the same time, such information is valuable in a number of technological fields, such as semiconductor technology, catalysis, lubrication, corrosion and its inhibition, metal finishing, adhesion, printing, and electrochemical technology. Vibrational spectroscopy, which has historically provided chemists, physicists, and engineers with incisive information about the chemical species present on surfaces, is one of the major surface analytical methods widely used today. This book describes a wide variety of vibrational methods and probes.

We begin with a brief summary of other widely employed surface analysis methods. A variety of surface measurement techniques are used for acquiring information on surface elemental composition, surface periodicity, surface electronic structure, surface morphology and defect maps, and the vibrational properties of surfaces and adsorbed species.[1-4] A reliable picture of the surface can be drawn from the combined use of these techniques. Auger electron spectroscopy is used mainly for the detection of surface impurities of light elements, low energy electron diffraction and grazing angle x ray diffraction are used for measurement of the surface periodicity, UV photoemission, electron energy-loss and metastable deexcitation spectroscopies for surface electronic characterization, x ray photoelectron spectroscopy for the oxidation state of surface atoms, and so on.

Most of the techniques employ electrons or ions as the incident or emitted particles. These particles interact strongly with materials, providing high sensitivity in the surface measurements.[5–8] Electron- and ion-based techniques, however, are normally confined to measurements at vacuum/solid interfaces and cannot generally be used for the technologically important *in situ* observation of surfaces in gaseous and liquid media. Furthermore, electrons used in the measurements have an energy sufficient to bring on chemical changes of the sample material present on the surface. The build-up of electrical charges due to the emission of electrons from excited surfaces may also result in a serious problem in the observation of phenomena occurring at the surface of dielectric materials.

Only a few of the electron- and ion-based techniques provide the most valuable chemical information related to the vibrational spectra of the probed materials. High-resolution electron energy-loss spectroscopy (HREELS)[9] is widely used for the observation of vibrational spectra of surface species of submonolayer coverages, but provides rather poor resolution compared to optical spectroscopies and is restricted to measurement in vacuum. Rapid transit of the sample from the reaction cell to the high vacuum chamber and the freezing of weekly adsorbed species are, therefore, used for *quasi-in-situ* observations. Inelastic tunneling spectroscopy is another electron-based vibrational spectroscopy.[10–12] Vibrational spectra of molecules adsorbed on an oxide or oxide-supported metal are measured in a metal-thin oxide layer tunneling junction. This method can detect adsorbates at monolayer or submonolayer coverages. However, the required presence of a metal-oxide-metal sandwich junction and liquid helium temperatures during the measurement are serious disadvantages for the *in situ* observation of surface species.

Scanning tunneling microscopy (STM)[13–17] provides information on surface topography and can be used in gaseous and liquid media. This and related techniques also reveal the electronic states of surface atoms and the distribution of work function, as well as the interatomic forces between surface atoms and atoms at the vertex of the probe tip, but as yet do not provide information about the vibration of surface species.

Optical methods using visible and near-ultraviolet light as the probe are used for the *in situ* observation of solid surfaces in various gaseous and liquid media. Ellipsometry[18,19] and reflection spectroscopy,[18,20] among others, have been applied extensively to the measurement of the optical properties of bulk solids and thin films. In particular, the former is used for the determination of the thickness of thin films on bulk substrates, and the latter is used for the *in situ* observation of electrode surfaces in combination with the potential modulation technique.

Optical vibrational spectroscopy or infrared and Raman spectroscopy is less sensitive than the counterpart (HREELS) of the electron-based techniques, but is applicable to the measurement of gas/solid, liquid/solid, and solid/solid interfaces. The wealth of characteristic group frequencies accumulated to date permits plausible chemical structures of surface species to be deduced from the measured spectra. This book deals with infrared reflection and emission spectroscopy, as well as normal and surface-enhanced Raman spectroscopy. Infrared transmission spectroscopy, which is used extensively for the observation of species adsorbed on finely divided metal particles, on porous oxides, and on porous silicon, has furnished us with invaluable information for acquiring insight into the mechanism of adsorption and catalysis. However, since excellent textbooks and review articles have been published on this methodology[21–26] and this book deals with the surfaces of bulk materials, infrared transmission spectroscopy is not included here. Also omitted for the scope of publication is an advanced theoretical discussion. The reader interested in this respect is referred to the excellent review articles cited in the following sections.

2. Infrared and Raman Observation of Chemical Species on Bulk Solid Substrates

Although photons are intrinsically less sensitive probes than electrons, they are used at the present time to detect species of a few monolayers or submonolayer coverages on bulk materials, taking full advantage of newly developed instruments and techniques. The infrared external and internal reflection methods are used for observing spectra of adsorbed films and surface layers of bulk materials. Infrared spectra of thin films on bulk metal substrates are readily recorded with the oblique incidence reflection (or reflection–absorption) method,[27] providing information about chemical species present, their orientations, and the film thickness. When the film thickness is a few micrometers or more, however, IR spectra of the film material, which is scraped from the substrates, are usually measured with the micro-KBr-pellet technique. Spectra of thin films of the same thickness on infrared–transparent semiconductor substrates are also obtained with the conventional transmission method. Evaporation of a thin Au or Ag film on top of the sample film may be effective for enhancing the signals of the film,[28] but one must remember that the enhancement occurs only in the direct vicinity of the metal film.

When solid surfaces are investigated in absorbing gaseous and liquid media, strong IR absorptions of the media interfere with the observation of faint signals of the surface species. The polarization modulation and related methods are effective for minimizing such interference, and are often applied to the *in situ* observation on bare and oxide-covered metal surfaces. The electrode potential modulation is a complementary technique of the polarization modulation method used in the *in situ* observation of electrode surfaces. It provides in real time the change in the IR spectra from surface species arising from the electrode potential sweep.

The wavelength modulation method involves the observation of derivative spectra and is higher in sensitivity than the polarization modulation in the measurement of sharp spectral features from surface species.[29] There exist, however, substantial disadvantages to this technique. The sensitivity of this technique decreases significantly when the IR features of the surface species are broad. The absorbing gaseous media result in enormous features in the recorded spectra, making the use of an evacuated spectrometer indispensable. This technique is recommended for the observation of sharp signals from species adsorbed on well-defined metal surfaces in ultrahigh vacuum.

The infrared internal reflection (or attenuated total reflection) method is employed for the observation of semiconductor surfaces and surface layers of polymers.[30] This method is applicable also to the measurement of thin films on plane metal and dielectric surfaces brought into contact with the internal reflection element. Its sensitivity is enhanced by multiple internal reflections, and spectra of adsorbates at submonolayer coverages on semiconductor surfaces are acquired by using the semiconductor itself as the internal reflection element. Most semiconductors, however, are transparent to infrared light in rather limited wavelength regions, so the internal reflection measurement using the semiconductor as the reflection element is often confined to a limited region. In the internal reflection method, the probing depth of the incident radiation changes depending upon the angle of incidence. The change in the spectrum with the distance from the surface may thus be observed nondestructively. Deposition of a Au or Ag island film onto the base plane of the internal reflection element is used as an effective technique for the observation of the outermost layer of polymeric materials.

The observation of vibrational spectra of species directly bound to the substrate surface is often crucial to understanding the mechanism of heterogeneous catalysis, adhesion, lubrication, corrosion inhibition, and epitaxial growth of thin films. In the observation of liquid/solid and solid/solid interfaces, however, surface IR and Raman techniques usually provide information about interphase layers of 100 nm or more in thickness in addition to recording the signals from the bulk. We are then obliged to measure difference spectra, subtracting appropriate reference spectra from the sample spectra, when information about molecules in direct contact with the substrate is desired. Any technique bringing about the enhancement of signals only from the molecules present at the interface should be advantageous. Although surface Raman and infrared spectra enhanced by a short-range "chemical effect"[31,32] seem to fulfill this requirement, the enhancement is specific to particular chemical systems. The measurement of "chemically" enhanced surface vibrational spectra is thus applicable only to limited systems.

The electromagnetic field excited at island films of coinage metals (Au, Ag and Cu) has a sizable strength only in the direct vicinity of the islands both in the visible and in the infrared. Consequently, infrared and Raman spectra of very thin interface layers can be obtained through the introduction of an island film of a coinage metal into the interfacial plane.

Infrared emission spectroscopy may be utilized for the investigation of solid surfaces.[33] No incident probe is needed in the emission measurement: it is a non-interfering measurement. Infrared emission has an appreciable intensity, particularly in the long wavelength region, when the temperature of the sample rises beyond room temperature. The emission intensity from species on metal substrates shows a sharp angular distribution, exhibiting a peak at a large observation angle. In addition, the emission is polarized perpendicular to the metal surface. Polarization modulation, therefore, is used to eliminate stray emission and a consequent enhancement in the signal-to-noise ratio of the measured spectrum.[34] In the observation of species on rough or curved metal surfaces, infrared emission measurements commonly provide better spectra than the reflection measurement.[33,35]

Infrared diffuse reflectance[20,36,37] and photo-acoustic[38] spectroscopies are usually employed for the measurement of species on rough solid substrates and on fine solid particles. The measurement of spectra with these techniques is greatly facilitated by the use of FTIR spectrometers. The photoacoustic measurements require the presence of gas or liquid phase over the sample. The scattered and reflected radiation from the sample gives no signal in the photo-acoustic observation, in contrast to the diffuse reflection measurement, and the photoacoustic measurement often gives better results than the diffuse reflection method.[39] In addition to the measurement of rough surfaces, photoacoustic IR spectroscopy has been applied to the semiquantitative analysis and the determination of the orientation of molecules in Langmuir–Blodgett films on flat polymer disks.[40] Spectra of surface layers of polymeric materials are measured by means of the IR photoacoustic method as well.[41,42]

Normal Raman spectroscopy is less sensitive than infrared spectroscopy, but can be used for the observation of surface species on metal and transparent solid substrates, provided the species have a large Raman cross section and the measurement is performed under the optimum experimental conditions. In a classical picture, Raman scattering of surface species can be separated into two processes: (1) the excitation induced by the electromagnetic field at the surface from the vibrational ground state in the electronic ground state to an excited virtual state; and (2) the subsequent transition from the excited state to an excited vibrational level in the electronic ground state accompanied by the spontaneous emission of Raman shifted light. The probability of the induced transition to the excited state increases with the strength of the electromagnetic field at the surface. It follows that the intensity of Raman scattered light increases with the enhancement of the surface electromagnetic field. The enhancement of the surface electromagnetic field is brought about on highly reflecting metal surfaces by the incidence of the exciting laser light at a large angle, and on the transparent solid surfaces, by the incidence of the laser beam through the solid substrates at the critical angle of total reflection.

The intensity of the scattered light changes significantly depending upon the collection (or observation) angle as a result of the reflection at the surface or interface and of the interference. On metal surfaces, the intensity of the scattered light reaches maximum at a large collection angle, and the collection of Raman shifted light through the solid substrate at an angle near the critical angle gives the strongest Raman signals for the species on the transparent substrate.

The electrode potential modulation method is utilized also in the Raman measurements for the *in situ* observation of species on metal and semiconductor electrode surfaces. Reaction intermediates and metastable species generated at the electrodes can be monitored *in situ* and in real time with this modulation technique by taking advantage of the intensity enhancement from the resonance or pre-resonance Raman scattering, when colored species are generated on the electrode.

The intensity of Raman lines increases enormously in surface-enhanced Raman scattering (SERS). The enhancement of Raman scattered light arises from electromagnetic and "chemical" mechanisms.[31] The incident laser light induces a surface plasmon polariton (SPP) and a collective electron resonance at rough surfaces of coinage metals. The strong electromagnetic field of the SPP or of the collective electron resonance induces the enhancement of the Raman shifted light and the Raman shifted light again induces SPP or the collective electron resonance, resulting in further intensity increase of the Raman scattered light. Surface plasmon polariton and collective electron resonance are also generated on coinage metal films in the internal reflection arrangement.

The "chemical" enhancement is generally believed to be based on the metal electron-mediated resonance Raman effect, which involves the charge transfer from metal states to affinity levels of the adsorbates. When both of the mechanisms are operating, an enhancement of Raman scattered light of six orders of magnitude may take place.

Surface-enhanced Raman scattering seems to be an excellent tool, especially for the *in situ* observation of electrode surfaces. However, there exist several serious problems in the application of this method. The strong electromagnetic field is generated only on the rough surfaces of a limited number of metals (i.e., Ag, Au, Cu and some other less common metals) using light in the wavelength region where the metal shows free-electron metal-like behavior. Several techniques have been developed to overcome this difficulty. For instance, a small quantity of a transition metal is deposited on SPP-supporting metal gratings for the observation of adsorption on the transition metal.[43]

In general, the enhancement factor of the electromagnetic mechanism is insufficient for the detection of surface species at submonolayer coverages. The "chemical" mechanism is often operating on the surfaces of SPP-supporting metals, resulting in intense Raman signals of species chemisorbed at the metal surfaces. The "chemical" mechanism, however, has chemical specificity, because it is a kind of the resonance Raman scattering. As a result, the metal surfaces normally cannot be surveyed with the SERS measurement.

The infrared–visible sum frequency generation method is a newly developed and promising technique for the *in situ* observation of flat surfaces and interfaces.[44] Sum frequency generation (SFG) is a second-order nonlinear optical process and is forbidden in a centrosymmetric medium. Such symmetry will necessarily be broken at an interface, and SFG is allowed at surfaces. When the phase on each side of the interface is centrosymmetric, SFG signals are generated only at an interface region of a few angstroms thickness. As a result, SFG is particularly surface-sensitive and can be used to elicit vibrational information from gas/solid, liquid/solid, and solid/solid interfaces unless the solid is non-centrosymmetric.

When sufficiently intense pulses of laser beams of frequencies ω_1 and ω_2 are incident on an interface, the output has a frequency $\omega_1 + \omega_2$ besides others. If ω_1 is in the infrared and ω_2 in the visible, the generated sum frequency falls in the visible region, where a highly efficient detector can be used. In the SFG measurement, ω_2 is constant and ω_1 is tuned in an infrared region. When ω_1 is tuned to a vibrational frequency of the surface species, the $\omega_1 + \omega_2$ output exhibits a resonance enhancement, providing the vibrational spectrum of the surface species. Sum frequency generation measurement has been applied to the investigation of species on water, glass, silica, metals, and semi-conductors.[45–47]

Since *s*- and *p*-polarized laser beams can be used as the incident light, four different polarization combinations are used in the measurement, giving rise to

Table 1. Suggested Optical Method of Measuring Vibrational Spectra of Thin Films on Bulk Substrates

---

Thin Films on Metal Substrates
  Plane Metal Surfaces
    Oblique incidence IR reflection
    IR polarization modulation
    IR internal reflection
    Infrared emission
    Sum frequency generation
    IR surface electromagnetic wave spectroscopy
    Raman scattering at the optimum condition of measurement
  Curved Metal Surfaces
    Infrared emission
    Raman scattering at the optimum measuring condition
    Sum frequency generation
  In situ Observation of Metal Surfaces in Absorbing Gaseous Medium
    IR polarization modulation
    IR internal reflection in the Kretschmann configuration
    Sum frequency generation
  Rough Metal Surfaces
    Infrared emission
    IR diffuse reflection
    IR photoacoustics
    Oblique incidence IR reflection
  Somewhat Thick Surface Films
    IR transmission measurement of micro-KBr-pellets of scraped film material
    Near-normal incidence IR reflection
    Backscattered Raman measurement

Surface Layers of Polymeric Materials
    IR internal reflection
    IR photoacoustics
    Total reflection Raman scattering

In-situ Observation of Metal Electrode Surfaces
    Electrochemical modulation IR and SNIFTIRS
    IR polarization modulation
    IR internal reflection in the Kretschmann configuration
    Surface-enhanced Raman scattering
    Sum frequency generation
    Potential modulation Raman (for the observation of electrochemically generated colored
        species)

Semiconductor Surfaces
    IR internal reflection

Table 1. *(Continued)*

Semiconductor Surfaces *(continued)*
    Sum frequency generation
    IR transmission (thick film)
    Total reflection Raman scattering (flat surface)
    Backscattered Raman measurement
    IR polarization modulation

Thin Films on Plane Dielectric Surfaces
    IR internal reflection
    Sum frequency generation
    Total reflection Raman scattering

---

the change in intensity of the SFG signals. The orientation of surface molecules, therefore, can be investigated by observing the polarization dependence of SFG spectra.[48] With picosecond laser pulses, time resolved measurements of SFG spectra can be performed for the investigation of surface reactions and surface dynamics.[49]

For a vibrational mode to be detected in the SFG measurement, it must be both Raman and infrared active.[47] A substantial background signal is superimposed on the signals from surface species in the SFG measurement of species on metals and semiconductors. However, the interference between these two types of signals provides information on the orientation of surface species.[47] A major disadvantage of this technique is the lack of tunable intense infrared laser sources which cover a sufficiently wide wavenumber region. The wide-region-covering laser source would also be useful in surface infrared spectroscopy, and when such a laser is available we expect to see further progress in surface science.

As mentioned above, each measuring method is efficient only under particular experimental conditions, and so, in the measurement of vibrational spectra of species on solid surfaces, an appropriate method should be chosen, taking into account the substrate material, the medium, the sample thickness, and the surface roughness. The solid surfaces are classified into nine categories, and the IR and Raman techniques applicable to each category are collected in Table 1. The basic principles and experimental techniques for each measuring method appear in the following chapters, except those of sum frequency generation, IR diffuse reflection, and IR photoacoustic spectroscopy, whose references[36–42,44–49] are listed at the end of this chapter.

References

1. P. F. Kane and G. B. Larabee, eds., *Characterization of Solid Surfaces,* Plenum, New York (1974).
2. A. W. Czanderna, ed., *Methods of Surface Analysis,* Elsevier, Amsterdam (1975).
3. N. Sheppard, *Spectrochim. Acta* Special Suppl., 149 (1989).
4. J. H. Block, A. M. Bradshaw, P. C. Gravelle, J. Haber, R. S. Hansen, M. W. Roberts, N. Sheppard and K. Tamaru, *Pure Appl. Chem.* **62,** 2297 (1990).
5. T. A. Carlson, *Photoelectron and Auger Spectroscopy,* Plenum, New York (1975).
6. H. Ibach, ed., *Electron Spectroscopy for Surface Analysis,* Springer, Berlin (1977).
7. B. Feuerbacher, B. Fitton, and R. F. Willis, eds., *Photo-emission and Electronic Properties of Surfaces,* Wiley, New York (1978).
8. C. R. Brundle and A. D. Baker, eds., *Electron Spectroscopy: Theory, Techniques and Applications,* Vols. 1–4, Academic, London (1977–1981).
9. H. Ibach and D. L. Mills, *Electron Energy Loss Spectroscopy and Surface Vibrations,* Academic, New York (1982).
10. T. Wolfram, ed., *Inelastic Electron Tunneling Spectroscopy,* Springer, Berlin (1978).
11. D. G. Walmsley, in *Vibrational Spectroscopy of Adsorbates* (R. F. Willis, ed.), Springer, Berlin (1980), p. 67.
12. P. K. Hansma, ed., *Tunnelling Spectroscopy,* Plenum, New York (1982).
13. G. Binnig and H. Rohrer, *Helv. Phys. Acta* **55,** 726 (1982).
14. Y. Kuk and P. J. Silverman, *Rev. Sci. Instrum.* **60,** 165 (1989).
15. A. V. Latyshev, A. L. Assev, A. B. Krasillnikov, and S. I. Stenin, *Surf. Sci.* **213,** 157 (1989).
16. R. J. Hamers, R. M. Tromp, and J. E. Demuth, *Phys. Rev. Lett.* **56,** 1972 (1986).
17. J. J. Boland, *Surf. Sci.* **261,** 17 (1992); *J. Vac. Sci. Technol.* **A10,** 2458 (1992).
18. R. F. Muller, ed., *Advances in Electrochemistry and Electrochemical Engineering,* Vol. 9, *Optical Techniques in Electrochemistry,* Wiley-Interscience, New York (1973).
19. R. M. A. Azzam and N. M. Bashara, *Ellipsometry and Polarized Light,* North-Holland, Amsterdam (1987).
20. W. W. Wendlandt and H. G. Hecht, *Reflectance Spectroscopy,* Wiley-Interscience, New York (1966).
21. L. H. Little, *Infrared Spectra of Adsorbed Species,* Academic, New York (1966).
22. M. L. Hair, *Infrared Spectroscopy in Surface Chemistry,* Dekker, New York (1967).
23. A. V. Kiselev and V. I. Lygin, *Infrared Spectra of Surface Compounds,* Wiley, New York (1975).
24a. J. T. Yates, Jr. and T. E. Madey, eds., *Methods of Surface Characterization,* Vol. 1, Plenum, New York (1987).
24b. A. T. Bell and M. L. Hair, eds., *Vibrational Spectroscopy for Adsorbed Species,* American Chemical Society, Washington, D.C. (1980).
25. N. Sheppard, *Ann. Rev. Phys. Chem.* **39,** 589 (1988); N. Sheppard and C. de la Cruz, *React. Kinet. Catal. Lett.* **35,** 21 (1987).
26. B. A. Morrow, in *Vibrational Spectroscopy for Adsorbed Species* (A. T. Bell and M. L. Hair, eds.), American Chemical Society, Washington, D.C. (1980), p. 119.
27. R. G. Greenler, *J. Chem. Phys.* **44,** 310 (1966).
28. A. Hartstein, J. R. Kirtley, and J. C. Tsang, *Phys. Rev. Lett.* **45,** 201 (1980).
29. P. Hollins and J. Pritchard, in *Vibrational Spectroscopy of Adsorbates* (R. F. Willis, ed.), Springer, Berlin (1980), p. 125.

30. N. J. Harrick, *Internal Reflection Spectroscopy,* Harrick Scientific Corp., Ossining, NY (1979).
31. A. Otto, *J. Raman Spectrosc.* **22,** 743 (1991).
32. J. P. Devlin and K. Consani, *J. Phys. Chem.* **85,** 2597 (1981).
33. D. Kember and N. Sheppard, *Appl. Spectrosc.* **29,** 496 (1975).
34. K. Wagatsuma, K. Monma, and W. Suëtaka, *Appl. Surf. Sci.* **7,** 281 (1981).
35. Y. Nagasawa and A. Ishitani, *Appl. Spectrosc.* **38,** 168 (1984).
36. G. Kortüm, *Reflectance Spectroscopy,* Springer, Berlin (1969).
37. M. Fuller and P. Griffiths, *Anal. Chem.* **50,** 6 (1980).
38. R. Rosencwaig, *Photoacoustics and Photoacoustic Spectroscopy,* Wiley, New York (1980).
39. D. W. Vidrine, *Appl. Spectrosc.* **34,** 319 (1980).
40. E. G. Chatzi, M. W. Urban, H. Ishida, J. L. Koenig, A. Laschewski, and H. Ringsdorf, *Langmuir* **4,** 846 (1988).
41. J. C. Donini and K. H. Michaelian, *Infrared Phys.* **24,** 157 (1984).
42. D. A. Saucy, S. J. Simko, and R. W. Linton, *Anal. Chem.* **57,** 871 (1985).
43. L-W. H. Leung and M. J. Weaver, *J. Am. Chem. Soc.* **109,** 5113 (1987).
44. Y. R. Shen, *Nature* **337,** 519 (1989).
45. X. D. Zhu, H. Suhr, and Y. R. Shen, *Phys. Rev.* **B35,** 3047 (1987); J. H. Hunt, P. Guyot-Sionnest, and Y. R. Shen, *Chem. Phys. Lett.* **133,** 189 (1987).
46. R. Superfine, P. Guyot-Sionnest, J. H. Hunt, C. T. Kao, and Y. R. Shen, *Surf. Sci.* **200,** L445 (1988).
47. A. L. Harris, C. E. D. Chidsey, N. J. Levinos, and D. N. Loiacono, *Chem. Phys. Lett.* **141,** 350 (1987).
48. P. Guyot-Sionnest, R. Superfine, J. H. Hunt, and Y. R. Shen, *Chem. Phys. Lett.* **144,** 1 (1988).
49. P. Guyot-Sionnest, P. Dumas, Y. J. Chabal, and G. S. Higashi, *Phys. Rev. Lett.* **64,** 2156 (1990); Y. J. Chabal, P. Dumas, P. Guyot-Sionnest, and G. S. Higashi, *Surf. Sci.* **246,** 524 (1991).

Infrared External Reflection Spectroscopy

1. Oblique Incidence Reflection Method

1.1. Background

Infrared spectroscopy has been widely employed for the investigation of species adsorbed on solid substrates.[1-3] However, because this technique was insufficient in sensitivity for observing weak signals of surface species, the use of a finely divided adsorbent was required to obtain a population of adsorbed species sufficient for generating good spectra. Although the finely divided system is of practical importance and is investigated widely by transmission and diffuse reflection IR spectroscopy, the parallel use of the *other technique* for characterizing the solid surfaces is difficult with this system. Thus, the need for a well characterized surface inevitably requires the use of bulk material.

Francis and Ellison were successful in observing infrared spectra of thin films on metal mirror surfaces by the external reflection method using high angles of incidence.[4] They also predicted theoretically and confirmed experimentally that, at the surface, the electric field is polarized normal to the metal surface except near normal incidence. Bartell and Churchill subsequently deduced in their polarimetric study that *p*-polarized light should give absorption spectra of species on metal surfaces with high sensitivity in the visible region as well.[5] Leja *et al.* and Hannah obtained, respectively, infrared spectra of xanthate adsorbed on copper mirrors and oxide films on aluminum using multiple reflection attachments.[6,7] However, most spectroscopists were reluctant to use this promising technique* until Greenler gave a quite convincing

---

*Absorption bands of films on semiconductor surfaces appear as increases in reflectivity at corresponding wavenumbers in the measured reflection spectra. Regarding films on metal surfaces, IR reflection spectra of inorganic films bear no resemblance to their absorption spectra: the bands shift to higher wavenumbers and are considerably deformed in shape. Although this reflection technique is generally called the reflection-absorption method, it is written in this volume as the oblique-incidence reflection method for these reasons.

elucidation of the increase in sensitivity at high angle incidence of p-polarized light.[8]

The increase in population of surface species irradiated by the infrared beam certainly gives rise to the increase in signal strength at high angles of incidence. If the incident angle is ϕ_1, the population increases in proportion to $\sec\phi_1$. Another important factor is the increase in strength of electric field at the metal surface. The intensity of the IR absorption band increases in proportion to the strength of the electric field inducing the absorption. In the conventional transmission measurement, the electric field of the incident IR wave induces absorptions. On a highly reflective metal surface, the incident IR wave combines with the reflected wave of approximately equal strength, producing a standing wave. It is this standing wave that induces IR absorption of species present on the metal surface.

The strength of the standing wave at the metal surface changes significantly depending upon the polarization state of the incident IR beam and the angle of incidence. This can be well understood by the approximation given by Greenler.[8] The phase shift upon reflection by a highly reflective metal for the light polarized perpendicular to the plane of incidence (s-polarized light) is close to $-\pi$, irrespective of the angle of incidence, as shown in Fig. 1. The amplitude of the standing wave, therefore, should be nearly equal to zero, as can be seen in Figs. 2a and 2b. In other words, a node of the standing wave forms at the metal surface.

Figure 1 also shows the phase shift for the light polarized parallel to the plane of incidence (p-polarized light), which changes from a value approaching zero at normal incidence to a value of $-\pi$ at grazing incidence. In consequence, the combination of the incident and the reflected beams generates a sizable electric field oscillating normal to the metal surface at high angles of incidence, as can be seen in Fig. 2d. In contrast to the normal field, the field parallel to the metal surface is very weak, independent of the incident angle of p-polarized light. In conclusion, the incidence of p-polarized light at a high angle is essential for obtaining a sizable electric field at the antinode of the standing wave at the metal surface and for obtaining the resulting increase of the absorption intensity of surface species.

The amplitude of the electric field oscillating normal to the metal surface is plotted in Fig. 3 versus incident angle of p-polarized light.[9] This figure shows that the amplitude reaches a maximum at a high incident angle of about 80°. Since the maximum amplitude is almost twice the magnitude of the incident IR light, the electric field on the metal surface is nearly four times as strong as the incident light. In addition, the population of surface species irradiated by the IR beam increases in inverse proportion to $\cos\phi_1$. As a result,

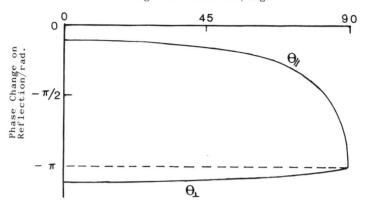

Figure 1. Phase shift of IR light reflected from a highly reflective metal surface. $\theta_{\parallel}$ = phase shift for *p*-polarized light, and $\theta_{\perp}$ = phase shift for *s*-polarized light [Reprinted from R. G. Greenler: *J. Chem. Phys.* **44**, 310 (1966)].

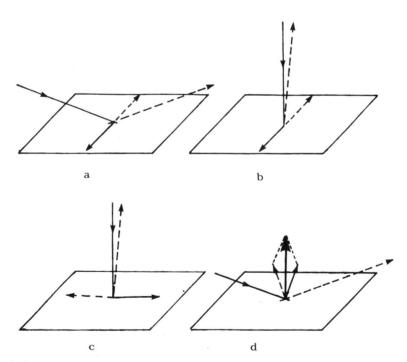

Figure 2. Combination of incident and reflected lights at metal surface. (a) *s*-polarized light, oblique incidence, (b) *s*-polarized light, near-normal incidence, (c) *p*-polarized light, near-normal incidence, (d) *p*-polarized light, oblique incidence.

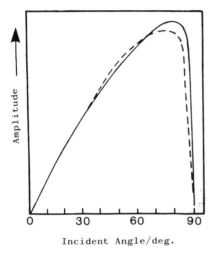

Figure 3. Amplitude of electric field oscillating normal to the metal surface versus angle of incidence. ——, Al (wavelength = 7 μm); – –, Ag (wavelength = 3 μm) (adapted from Ref. 9).

the signal strength of surface species shows a peak at a high incident angle of about 88° on highly reflective metal surfaces.[8] This optimum angle of incidence is slightly smaller than the pseudo-Brewster angle of the substrate metal.[10] When the change in surface electric field by the introduction of the film can be neglected, the signal strength in the reflection spectrum should be more than 50 times greater than in the transmission measurement at normal incidence of the same sample layer separated from the metal substrate.[11] The surface field change for the introduction of the film will be mentioned below.

    In the reflection measurement the electric field oscillating normal to the metal surface is sizable and the field parallel to the surface is negligible. It follows that, whereas the components normal to the surface of vibrating dipoles of surface species can interact with the surface electric field, giving rise to absorption bands, the parallel components produce absorption of negligible intensity. This is one of two effects that engender the surface selection rule. A vibrating dipole of surface species is accompanied by an image dipole within the metal. The surface selection rule derives also from the image dipole. Whereas a dipole vibrating normal to the surface gives rise to an image dipole which is parallel to the dipole of the surface species, the dipole vibrating parallel to the metal surface generates an image dipole which is oriented antiparallel, resulting in the compensation of the dynamic dipole moment.

When a metal particle is of sufficient size to generate an image dipole, this selection rule should hold true for species adsorbed on the metal particle.[12] Greenler *et al.* calculated that the selection rule should apply to species adsorbed on metal particles larger than about 2 nm in diameter.[13] The surface selection rule is often applied to determine the orientation of species on metal surfaces. Typical examples will be shown later.

For a quantitative treatment of the reflection spectra of thin films on solid surfaces, one must consider a three-layer model, taking into account the optical properties of the thin film. McIntyre and Aspnes formulated a linear-approximation theory which has greatly simplified the theoretical treatment of the complex system.[14] The linear-approximation theory is summarized below.

Suppose that a collimated beam of IR light of wavelength λ (in vacuum) is incident from the transparent medium at an angle ϕ_1 on a plane, parallel-sided, homogeneous, isotropic film of thickness d supported on a solid substrate, which is assumed homogeneous and isotropic (Fig. 4).

Using I_i and I_r for the intensity of the incident and the reflected light, respectively, the reflectance R is given by

$$R = I_r/I_i \tag{1}$$

The quantities ε_1, $\hat{\varepsilon}_2 (= \varepsilon_2' - i\varepsilon_2'')$, $\hat{\varepsilon}_3 (= \varepsilon_3' - i\varepsilon_3'')$ in Fig. 4 are dielectric constants of the transparent medium, the thin film, and the solid substrate, respectively. Since the thin film and the substrate are assumed absorbing, the dielectric

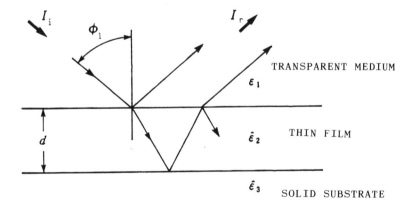

Figure 4. Reflection from a solid substrate covered with a thin film.

constants of these phases are complex, and ε' and ε'' are the real and the imaginary parts of the complex dielectric constant $\hat{\varepsilon}$.

The Fresnel reflection coefficient, r_d, of the three-phase system shown in this figure for $v(s$ or $p)$ – polarized light[15] is given by

$$r_{dv} = \frac{r_{v12} + r_{v23} \exp(-2i\delta)}{1 + r_{v12} + r_{v23} \exp(-2i\delta)} \qquad (2)$$

$$v = p, s$$

where δ is the change in phase and amplitude of the beam traversing the absorbing film and is given by

$$\delta = 2\pi\,\hat{n}_2\,d\cos\phi_2/\lambda \qquad (3)$$

The Fresnel reflection coefficients of the j/k interface for s- and p-polarized light are given by

$$r_{sjk} = \frac{\mu_k\hat{n}_j\cos\phi_j - \mu_j\hat{n}_k\cos\phi_k}{\mu_k\hat{n}_j\cos\phi_j + \mu_j\hat{n}_k\cos\phi_k} \qquad (4)$$

$$r_{pjk} = \frac{\hat{\varepsilon}_k\hat{n}_j\cos\phi_j - \hat{\varepsilon}_j\hat{n}_k\cos\phi_k}{\hat{\varepsilon}_k\hat{n}_j\cos\phi_j + \hat{\varepsilon}_j\hat{n}_k\cos\phi_k} \qquad (5)$$

where μ_j is magnetic permeability of the material j and is related to the complex refractive index $\hat{n}_j$ via

$$(\mu_j\,\hat{\varepsilon}_j)^{1/2} = \hat{n}_j = n_j - ik_j$$

where n_j and k_j are, respectively, the refractive index and the extinction coefficient of the material j. The magnetic permeability can be taken as unity in the optical frequency range. The angle of incidence, ϕ_j, and the angle of refraction, ϕ_k, are complex if the corresponding materials are absorbing, and are interrelated by Snell's law:

$$\hat{n}_j\sin\phi_j = \hat{n}_k\sin\phi_k$$

The theoretical equation for the reflectance, R_d $(= |r_d|^2)$ of the system shown in Fig. 4 is complicated. One cannot gain any physical insight into the change in reflection spectra depending upon the thickness and the optical constants of the film from this complicated equation. If the thickness of the surface film is less than ~25 nm, however, d/λ is much smaller than unity, because the wavelength of the IR light generally used in the experiments falls within a range of 2.5–25 μm. Under conditions of $d/\lambda \ll 1$, we can use a linear approximation when expanding Eq. (2) in powers of δ. Writing R_0 for the reflectance of the film-free system and $\Delta R (= R_d - R_0)$ for the increase in reflectance arising from the introduction of the film, the normalized reflectance change, $\Delta R/R_0$, can be written for s- and p-polarized light, respectively, as

$$\left(\frac{\Delta R}{R_0}\right)_s = \frac{8\pi \, dn_1 \cos \phi_1}{\lambda} \, Im \left(\frac{\hat{\varepsilon}_2 - \hat{\varepsilon}_3}{\varepsilon_1 - \hat{\varepsilon}_3}\right) \tag{6}$$

$$\left(\frac{\Delta R}{R_0}\right)_p = \frac{8\pi \, dn_1 \cos \phi_1}{\lambda} \, Im \left[\left(\frac{\hat{\varepsilon}_2 - \hat{\varepsilon}_3}{\varepsilon_1 - \hat{\varepsilon}_3}\right) \times \right.$$

$$\left. \left\{\frac{1 - (\varepsilon_1/\hat{\varepsilon}_2\hat{\varepsilon}_3)(\hat{\varepsilon}_2 + \hat{\varepsilon}_3)\sin^2 \phi_1}{1 - (1/\hat{\varepsilon}_3)(\varepsilon_1 + \hat{\varepsilon}_3)\sin^2 \phi_1}\right\}\right] \tag{7}$$

where Im designates the imaginary parts of the complex quantities.

When the substrate is a highly reflective metal and the film is weakly absorbing (i.e., $\hat{\varepsilon}_3 \gg \varepsilon_1$ and $\hat{\varepsilon}_3 \gg \hat{\varepsilon}_2$), ε_1 and $\hat{\varepsilon}_2$ can be neglected against $\hat{\varepsilon}_3$. Equation (6) retains no optical constants of the film, showing that incident s-polarized IR light should give no spectrum at any angle of incidence. For incident p-polarized light, Eq. (7) can be rewritten to a good approximation (unless $\phi_1 \simeq 90°$) as

$$\left(\frac{\Delta R}{R_0}\right)_p = \frac{-16\pi n_1^3 n_2 k_2 d \sin^2 \phi_1}{(n_2^2 + k_2^2)^2 \lambda \cos \phi_1} \tag{8}$$

or

$$\left(\frac{\Delta R}{R_0}\right)_p = \frac{-4n_1^3 n_2 \sin^2 \phi_1}{(n_2^2 + k_2^2)^2 \cos \phi_1} \, \alpha d \tag{9}$$

where $\alpha \ (= 4\pi k_2/\lambda)$ is the absorption coefficient of the film. For an organic film on a highly reflective metal in vacuum or air, Eq. (9) can be further simplified to yield Eq. (10), because usually $n^2 \gg k^2$,[16]

$$\left(\frac{\Delta R}{R_0}\right)_p = \frac{-4\sin^2 \phi_1}{n_2^3 \cos \phi_1} \, \alpha d \tag{10}$$

Equation (10) implies that an approximate IR absorption spectrum of the film is obtained with incident p-polarized light at a high angle as long as the anomalous dispersion of the refractive index is small.

If the infrared absorption of the thin film isolated from the metal substrate is measured by the conventional transmission method at normal incidence, the intensity I_d of IR light after passing through the film of thickness d is expressed as

$$I_d = I_0 \exp(-4\pi k_2 d/\lambda) \tag{11}$$

where I_0 is the intensity of the incident light. When $d/\lambda \ll 1$, to a good approximation Eq. (11) can be reduced to

$$I_d = I_0(1 - 4\pi k_2 d/\lambda)$$

Then we have the normalized intensity change, $\Delta I/I_0$, in the simple form of

$$\Delta I/I_0 \equiv (I_d - I_0)/I_0 = -4\pi k_2 d/\lambda = -\alpha d \tag{12}$$

A comparison of Eq. (10) with Eq. (12) indicates that the sensitivity in the reflection method is $4\sin^2 \phi_1/n_2^3 \cos \phi_1$ times greater than in the transmission method. The quantity $(4\sin^2 \phi_1/n_2^2)$ in the enhancement factor is the electric field strength within the surface film generated upon incidence of the light of unit strength, and the remaining quantity in the enhancement factor $(1/\cos \phi_1)$ is the area of film sampled by the light beam of unit cross section area.

Equations (9) and (10) show that, when the absorption coefficient α remains constant regardless of film thickness, the absolute value of the normalized reflectance change, $|\Delta R/R_0|_p$, increases in proportion to the film thickness d. This relation is useful for the estimation of thickness of surface films on metal substrates, as long as $d/\lambda \ll 1$.[17] Attention must be paid, however, because the absorption coefficient may change with film thickness. The coefficient of chemisorbed species may change with coverage.[18] The first layer of a Langmuir–Blodgett film may be different in molecular orientation from the outer layers, resulting in the change of the coefficient with film thickness.

The electric field of the standing wave decreases gradually in intensity from the antinode at the metal surface to the node in the medium. When the condition $d/\lambda \ll 1$ is fulfilled, the electric field has a sizable strength at the film/medium interface. In consequence, the adsorption at the surface of semiconductor, polymer, or oxide catalyst may be observed by the reflection measurement, provided a thin film of these materials is formed on a surface of highly reflective metal. At the same time, the observation of this system may provide us with information about the change in the film induced by the adsorption. Several examples of the observation will be shown in subsection 2.1.3. of this chapter.

Bermudez computed the electric field intensity at the surface of a silicon film on nickel.[19] The surface-normal component of the p-polarized electric field is shown in Fig. 5. This figure shows that the intensity of the normal component decreases slightly in the presence of a film of thickness less than 50

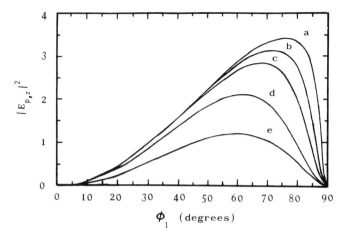

Figure 5. Intensity of surface-normal component of the *p*-polarized electric field (for unit incident field intensity) versus incident angle (ϕ_1) for a silicon film on a nickel mirror as a function of film thickness. $\lambda = 6.1$ μm (a) Ni (b) 50 nm Si/Ni (c) 100 nm Si/Ni (d) 200 nm Si/Ni (e) Si [Reprinted from V. M. Bermudez, *J. Vac. Sci. Technol.* **A10**, 152 (1992)].

nm. The electric field parallel to the surface can not be neglected at the thin film surface, but its intensity is more than one order of magnitude smaller than the normal field. The sensitivity, or normalized reflectance change $(\delta R/R_0)_p$, decreases rapidly with increase in film thickness, as shown in Fig. 6, and this sensitivity decrease is strongly wavelength-dependent.[19] This figure suggests that a condition where $d/\lambda^0 \leq 0.01$ (λ^0 is the shortest wavelength of the investigating range) is required for observing species on the film surface at high sensitivity.

The reflection method may be applied to observe surface species on a transparent solid substrate. However, when the solid material has a high refractive index, as in the case of semiconductors (Ge, Si, GaAs), the internal reflection method is higher in sensitivity than the external reflection method. Furthermore, the sensitivity of the internal reflection method can be enhanced by multiple reflection. The use of the internal reflection measurement is therefore preferable for the observation of the surface species on solid transparent materials having high refractive index. The external reflection method, nevertheless, is still useful for the observation of the surfaces of semiconductors at elevated temperatures where internal reflection observation is hampered by free carrier absorption and the surfaces of materials very weakly absorbing in a wide wavelength range.

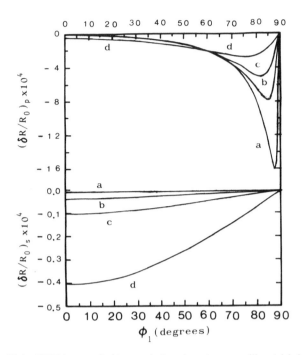

Figure 6. Sensitivity ($\delta R/R_0$) versus incident angle for adsorption on a silicon/nickel mirror as a function of silicon film thickness. The adsorbate is modeled as a 3-Å-thick layer of water ($\hat{n} = 1.311 - 0.132i$ at the adsorption maximum of 6.1 μm); (a) Ni (b) 50 nm Si/Ni (c) 100 nm Si/Ni (d) 200 nm Si/Ni. The sensitivity ($\delta R/R_0$) is slightly different from $\Delta R/R_0$ in the text, because $\delta R \equiv R_d - R^0$, where R^0 is the value of reflectance calculated assuming that the adsorbate is nonabsorbing [Reprinted from V. M. Bermudez, *J. Vac. Sci. Technol.* **A10**, 152 (1992)].

The sensitivity of the external reflection method was calculated for species on a silicon surface.[19,20] For a transparent substrate, R_0 is strongly dependent on ϕ_1 and falls to zero upon incidence of *p*-polarized light at the Brewster angle. The use of ΔR instead of $\Delta R/R_0$ is therefore appropriate for determining the sensitivity. ΔR_s and ΔR_p of a poly(vinylacetate) film ($d = 80$ nm)/silicon system are plotted against incident angle in Fig. 7.[20] Bermudez recently computed δR of a system of H_2O/Si.[19] The results, as shown in this figure, are in perfect qualitative agreement with his data. This figure shows that the highest sensitivity is obtained with the incidence of *p*-polarized light at a high angle near 85°. Considering that the noise level is equivalent to $\Delta R/R_0 = 1 \sim 5 \times 10^{-4}$ in most reflection measurements, this method can be applied to the observation of

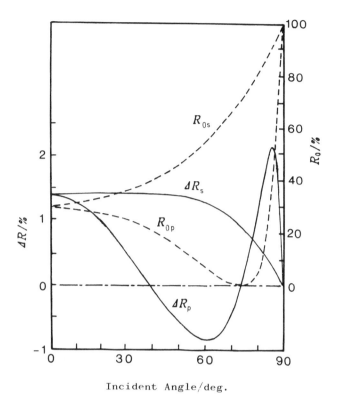

Figure 7. Reflectivity of bare silicon ($n = 3.42 - 0.00i$) at a wavelength of 5.7 μm and reflectivity change ΔR arising from the presence of a poly(vinylacetate) film ($\hat{n} = 1.4 - 0.51i$) on silicon for p- and s-polarized light (adapted from Ref. 20).

moderately absorbing surface films thicker than 10 nm on semiconductors.[21] Figure 7 also shows that absorption by a surface film on a transparent solid yields an increase in reflectivity except at an incidence of p-polarized light at around 60°, which gives an "inverse absorption." It should be noted that the strict surface selection rule cannot be applied to vibrations of species on transparent substrates, even if p-polarized light is incident at a high angle, because the electric field oscillating parallel to the surface is weak but has an appreciable strength in comparison to the normal field.[19]

External reflection spectroscopy can also be used for the observation of a thin film on a polymer substrate. When the reflectance of the film-covered substrate for p-polarized light (R_p) and that of the bare substrate (R_{p0}) are measured at the Brewster angle of the substrate, the quantity $-\log(R_p/R_{p0})$ gives relatively undistorted spectra for the film.[22]

1.2. Instruments

For the measurement of the reflection spectra, either a dispersive spectrometer or an interferometer is used. Interferometers have signal-to-noise ratio and spectral resolution superior to dispersive spectrometers in conventional measurement of bulk samples.[23] However, there are significant differences between the conventional and the surface measurements. In the surface measurement, the background photon noise may limit the signal-to-noise ratio, and the throughput of the optical system may be limited by the use of small samples, resulting in a decrease in the Fellgett and the Jacquinot's advantages of interferometers. A comparison of Fourier transform and dispersive methods in surface measurement was made,[24,25] and the conclusions drawn from that comparison are briefly given here. Generally speaking, Fourier transform may be the method of choice when spectral resolution or information of a wide spectral region are required. Dispersive spectrometers may be better for the observation of a narrow spectral region. In the observation of a narrow region, the application of polarization modulation is preferable, because the modulation brings about an improvement of the signal-to-noise ratio of the dispersive method as well as the sufficient cancellation of background absorption and emission from the sample at moderately elevated temperatures, as described in the next section.

Because the background spectrum may give rise to serious problems, the evacuation of the entire optical system is often applied. The atmospheric absorptions arising from the presence of H_2O and CO_2 may be adequately reduced by evacuating the whole system, despite technical difficulties that may occur as a result. For example, the possibility of electric discharge increases between terminals in the evacuated system. The purging of the optical path with dry air or nitrogen has been commonly used; however, the partial pressure of residual gas is higher than in the evacuation method and the drift of the pressure may reduce the signal-to-noise ratio of the recorded spectra. The double beam operation of spectrometers presents few technical difficulties and brings about a high level of cancellation of the background spectrum. It may also be used in the *in situ* observation of solid surfaces in absorbing reactant gas. An example of a double-beam FTIR spectrometer is shown in Fig. 8.[26] Both the spectrometer (Mattson, Cygnus 100) and the external optical bench are purged with dry nitrogen. Almost complete cancellation of the background absorption of H_2O and CO_2 was achieved with this apparatus. Figure 9 shows a schematic drawing of a UHV

FTIR SPECTROMETER

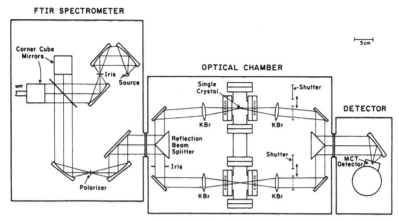

Figure 8. Schematic drawing of the optical arrangement of the double beam FTIR spectrometer for surface research. The entire optical system is purged continuously with dry nitrogen [Reprinted from Zhi Xu and J. T. Yates, Jr., *J. Vac. Sci. Technol.* **A8**, 3666 (1990)].

chamber coupled with a double-beam dispersive spectrometer.[27] A Perkin-Elmer PE580B spectrometer is equipped with a Globar source and a thermo-couple detector. Absorption bands of oxygen adsorbed on silver were observed with this system in a wavenumber region of 300–400 cm^{-1}, where water vapor has strong absorptions. The external mirrors and the optics in the spectrometer are enclosed in an airtight box that is purged by dry air or nitrogen.

IR spectrometers are generally equipped with a thermal detector operating at room temperature (thermocouple, Golay cell, and pyroelectric sensor). Thermal detectors have uniform sensitivity over a wide wavelength region, but are low in sensitivity and slow in response. Pyroelectric sensors are rather fast in response and are used as the detector in interferometers, despite their low sensitivity. In general, they are not used in surface IR spectroscopy. Liquid nitrogen cooled InSb and HgCdTe are widely used as detectors in surface IR investigation. InSb has high sensitivity but can be used only in the wavenumber region above 1820 cm^{-1}. HgCdTe is noisier than InSb but can be used in regions as low as 400 cm^{-1}; this is because the band gap of HgCdTe can be changed by varying the compositions of HgTe and CdTe. An InSb/HgCdTe dual detector is often chosen for dispersive spectrometers. For observation in lower wavenumber regions, Ge:Cu and Si:Sb are used down to 330 cm^{-1}[28] and Ge:Ga to 83 cm^{-1}.[10]

Normally, a thermal emitter (Globar or Nernst glower) is used as the light source in the spectrometer. In surface IR measurement, the Globar (silicon

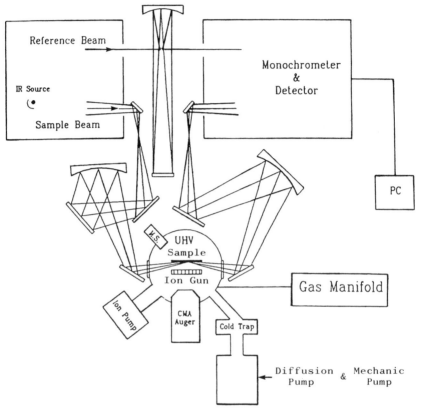

Figure 9. Schematic diagram showing the vacuum apparatus and the double beam spectrometer for the observation of metal surfaces [Reprinted from X. D. Wang, W. T. Tysoe, R. G.Greenler, and K. Truszkowska, *Surf. Sci.* **257**, 335 (1991)].

carbide) is often preferred to the Nernst glower (a mixture of zirconium, yttrium, and erbium oxides) because of its higher stability. The thermal emitter or the quasi-black body emitter radiates electromagnetic waves of energy $\rho d\lambda$ per unit area and unit time at a wavelength λ in an interval $d\lambda$, in conformity with Planck's radiation law:

$$\rho d\lambda = 2\pi c^2 h \lambda^{-5} \varepsilon(\lambda) [\exp(hc/\lambda kT) - 1]^{-1} d\lambda \qquad (13)$$

where T is the source temperature and $\varepsilon(\lambda)$ is the emissivity.[10] Equation (13) shows that the emitted power decreases sharply with an increase in wavelength and that the brightness is low in the far infrared region. When λ is large, a linear approximation can be applied to Eq. (13) after the expansion in powers of $hc/\lambda kT$. We then have

$$\rho d\lambda = 2\pi \, c\varepsilon \, (\lambda)kT\lambda^{-4} \, d\lambda \qquad (14)$$

Equation (14) shows that the emitting energy increases in proportion to the source temperature in the far infrared region. The increase in emitted energy upon a rise in temperature, however, is larger in the shorter wavelength region than in the far infrared, resulting in an increase in stray light intensity. The IR measurement in the long wavelength region, therefore, is not easy with a thermal source. A promising alternative is synchrotron radiation, which has higher brightness than the thermal emitter in the far infrared region.[29] The observation of surface species in the far infrared region has been successfully carried out using synchrotron radiation.[30,31] However, access to such the sources is not readily available to the general user.

The tunable infrared laser is another choice for the light source. Tunable lasers provide a very strong light source, making high resolution measurement possible.[32] However, no infrared laser that is continuously tunable over a wide wavelength region (2.5–25 μm) is available at present. Since each diode laser is tunable over a spectral range of 200–300 cm^{-1} by varying its temperature, several diodes must be used in combination for the measurement of a wide wavelength region. Butler *et al.* used an array of four diode lasers as the radiation source and observed the vibrational spectrum of oxygen chemisorbed on aluminum through the use of polarization modulation.[33] The combination of tunable diode lasers and polarization modulation was also used in the investigation of adsorbed pyridine on oxide-covered aluminum.[34]

The reflection method is used in the investigation of the clean and well defined metal surfaces as well as of the "real" surfaces of materials in air or in reactant gas. The vibrational spectra of species on clean surfaces are observed most widely with high resolution electron energy loss spectroscopy (EELS), which has a sensitivity more than one order of magnitude greater than IR and permits rapid observation of a wide wavelength region (2.5–50 μm). However, the low resolution of EELS (≥ 20 cm^{-1}) has limited its use for the line shape, peak shift, and peak position determination. Hence, reflection IR spectroscopy is used for the investigation of adsorbate interactions, adsorption site, surface diffusion, structure of adsorbate layer, and other phenomena. Because parallel observation with the other surface-sensitive techniques is required in the investigation of clean surfaces, various types of configurations have been designed for the IR spectrometer-UHV chamber system. Several examples of the full configuration of a UHV chamber plus a spectrometer are shown in review articles.[10,35] An apparatus employed in the study of time-resolved IR spectro-scopy is shown in Fig. 10.[36] An FTIR spectrometer is coupled with an AES, an LEED, a mass spectrometer, and a thrice differentially pumped supersonic molecular beam doser. The molecular beam doser introduces a uniform flux of the sample gas in a "single-shot" mode.

The sensitivity of oblique incidence reflection spectroscopy is not quite sufficient for the investigation of various species adsorbed on solid substrates. Multiple reflections could enhance the sensitivity. However, the background reflectance of metals decreases upon incidence of p-polarized light at high angles, and a decrease of about 30% was calculated by Greenler for the optimum angle incidence.[37] The intensity of IR light, therefore, decreases rapidly with the increase in the number of reflections. Greenler calculated the optimum number of reflections, which can be obtained from the requirement for maximizing ΔR and which changes depending upon the angle of incidence. The obtained optimum number is rather small (2–6) if the IR beam is incident at the optimum angle. In many cases, one reflection may yield a band intensity

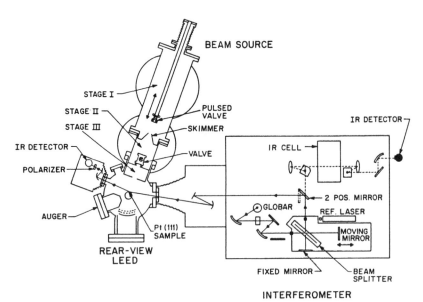

Figure 10. Schematic drawing of an apparatus used for time-resolved measurement. Principal components of the apparatus are a beam source, an UHV system with surface analytics, a rapid scanning FTIR spectrometer, and an InSb detector [Reprinted from J. E. Reutt-Robey, D. J. Doren, Y. J.Chabal, and S. B. Christman, *J. Chem. Phys.* **93**, 9113 (1990)].

acceptably close to the maximum intensity.[37] Furthermore, a large substrate is required for the multiple reflections, because the light is incident at a high angle. Hence multiple reflections are rarely applied, particularly to the investigation of adsorption on single crystal surfaces in UHV.

Various commercial attachments are available for the reflection measurement in air. Schematic drawings of the attachments are shown in Fig. 11. When the investigation of relatively thick films is required, the use of an attachment

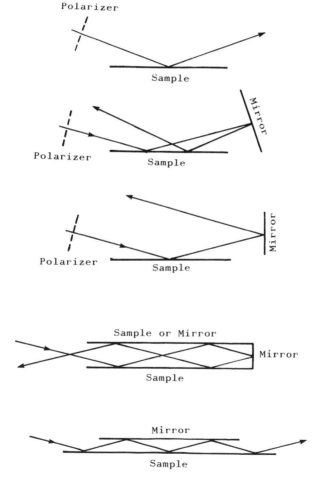

Figure 11. Schematic drawing of various attachments for the reflection measurement in air.

of variable incident angle is preferable as discussed in the next section. Grazing angle FTIR microscopy enables us to investigate small areas of metal surfaces. Reffner obtained spectra from 100-μm-diameter thin films of polycarbonate residue and solder flux residue on metal using a modified Cassegrain-type reflecting lens providing a maximum incident angle of 85°.[38]

The use of a polarizer increases the absorption intensity $(\Delta R/R_0)_p$, but the background decreases in its intensity at the same time; a wire grid polarizer reduces the intensity of IR light to about one-third of the incident intensity. When a dispersive spectrometer is employed, therefore, measurement without the polarizer may be preferable for the observation of rough surfaces and for multiple reflection observations.

The use of samples having a mirror surface is important for obtaining good reflection spectra. However, IR measurement of an unpolished surface is often required in practical work. Since the wavelength of infrared light is longer than for visible light, fairly good reflection spectra are obtained from samples having dull surfaces so long as the surface shows the sheen-effect[39] i.e., the dull surface behaves as a mirror, though imperfect, to the light incident at grazing angle. Spectra of lubricants and surface-finishing agents on thin metal wires can be obtained with the reflection method, if many pieces of the wire are stuck perpendicular to the incident plane and in contact with each other on the flat surface of a support to form a reflecting plane.

1.3. Analysis of Distorted Reflection Spectra

It is sometimes found that, when the IR spectra of isotropic films on metal surfaces are recorded with the IR reflection method, the spectra obtained bear a close resemblance to IR absorption spectra of the film materials. However, this is the case only when all the conditions of $d/\lambda \ll 1$, $\hat{\varepsilon}_3 \gg \varepsilon_1$, $\hat{\varepsilon}_3 \gg \hat{\varepsilon}_2$, $n_2^2 \gg k_2^2$, and faint anomalous dispersion of n_2 are fulfilled, as described in Section 1.1 of this chapter. These conditions can be fulfilled only in a system with an adsorbed layer (or a very thin organic film) on the surface of a highly reflective metal in vacuum or in air. The spectra of films on solid substrates obtained with the reflection method may be distorted and will not bear much resemblance to the absorption spectra of the film materials. Since in the investigation of "engineering" or "practical" surfaces, one frequently meets distorted spectra of surface films, the cause of the distortion and the analysis of the distorted spectra will be discussed in the following section.

If films on metal substrates have strong and broad absorption bands, the reflection spectra obtained will differ considerably from the absorption spectra of the film materials.[40,41] Since inorganic compounds generally have strong and broad IR absorptions, the features in reflection spectra of inorganic films are generally distorted in shape and are shifted to higher wavenumbers than in absorption spectra. A large shift exceeding 100 cm⁻¹ is sometimes observed. On transparent substrates, absorption bands of thin surface films bring about increases in reflectance at the frequencies of absorption instead of a reflectance decrease, as mentioned in the preceding section. The Kramers–Kronig (K–K) relation[42] is often useful for the analysis of the distorted reflection spectra,[43,44a] and elaborate analyses of deformed reflection spectra utilizing the K–K transformation appeared in literature recently.[44b,c]

The IR absorption spectrum of a thin amorphous silicon nitride (SiN_x) film photochemically deposited on a KBr plate shows a strong absorption band arising from the Si–N stretching vibration at 820 cm⁻¹, as shown in Fig. 12. Thin films of SiN_x deposited on gold show no bands in the region 700–900 cm⁻¹, but a strong feature appears at 1030 cm⁻¹ in reflection spectra, when observed *in situ* in the course of photochemical vapor deposition.[45] These two features differ in peak frequency so much from each other that they cannot be assigned to the identical vibrational mode without convincing reasoning. The disappearance of the Si–N band from the region of 700–900 cm⁻¹ may arise from the orientation of the Si–N bonds parallel to the metal surface, and the

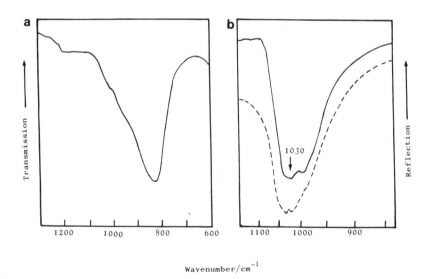

Figure 12. (a) Infrared absorption and (b) reflection spectra of thin SiN_x films. Angle of incidence in reflection measurement: 80°. ——, observed spectra; – – , calculated reflection spectrum from spectrum (a) for a film thickness of 20 nm [Reprinted from T. Wadayama and W. Suëtaka, *Surf. Sci.* **218**, L490 (1989)].

feature at 1030 cm^{-1} may stem from the Si–O stretching vibration of an oxidized layer of the film. However, such an orientation of the Si–N bond is unlikely, because the film is amorphous. The Si–O vibration gives rise to a feature around 1200 cm^{-1} in the reflection spectra. The feature at 1030 cm^{-1} might be a distorted band of the Si–N stretching vibration.

The Kramers–Kronig relation is useful for the analysis of distorted features. When the extinction coefficients, $k(\nu)$, of the film material are available, the values of refractive constants $n(\nu)$ are calculated by a Kramers–Kronig type relation:

$$n(\nu_0) - n(\infty) = \frac{2}{\pi} \int_0^\infty \frac{\nu k(\nu)}{\pi^2 - \pi_0^2} \, d\nu \qquad (15)$$

A difficulty lies in the practical application of this relation. While the k values are available within a limited spectral region, the integral of Eq. (15) requires the values for the entire region. Hence several approximations are used.[43,46] When an absorption band or a group of bands is isolated from the other bands, Eq. (15) may be written as[47]

$$n(\nu_0) - n'_\nu = \frac{2}{\pi} \int_{band} \frac{\nu k(\nu)}{\nu^2 - \nu_0^2} \, d\nu \qquad (16)$$

where the integration needs be taken only over the absorption band or the group of bands. The refractive index due to all other transitions, n_ν', can be estimated by an extrapolation from the visible region, because most of the contribution to refractive index comes from electronic transitions. Ohta and Ishida compared several numerical integration methods for the foregoing Kramers–Kronig transformation and concluded that, while Maclaurin's formula gives the most accurate results, the successive Fourier transform is less accurate than the other methods.[48]

The values of $k(\nu)$ were evaluated from the absorption spectrum of the SiN$_x$ film (thickness: = 130 nm) shown in Fig. 12a. From the values of $n(\nu)$ and $k(\nu)$, the reflection spectrum of a SiN$_x$ film of a given thickness on gold can be calculated by the use of Eq. (2), because the optical constants of gold are available from the literature. The calculated reflection spectrum of a SiN$_x$ film of 20 nm thickness is shown with a broken line in Fig. 12b. Fairly good agreement in peak frequency and band shape is obtained between the calculated and the observed spectra. The feature at 1030 cm^{-1} in the reflection spectrum is thus assignable to the Si–N stretching vibration.

Apart from very strongly absorbing bands, approximation Eqs. (9) and (10) can safely be applied to the bands when the film is thinner than about 0.01 λ_m (λ_m is the wavelength at the absorption peak). Figure 13 shows that the intensity, $|\Delta R/R_0|_p$, increases in proportion to the film thickness, in agreement with the above-mentioned equations.[17] However, the intensity of weak bands increases linearly to a point of greater thickness than does the intensity of strong bands, because the approximation equations are obtained by the expansion of the Fresnel coefficient r_d, to terms of first order in δ [$= 2\pi (n_2 - ik_2)d\cos \phi_2/\lambda$], bearing the extinction coefficient of the band in addition to the film thickness in the numerator. In consequence, when spectra of moderately thin films are recorded, a reduction in the relative intensity of strong bands may be seen even if the film is isotropic. When the film thickness increases further, strong bands begin to change shape. For example, in the spectra of polymer films having a thickness of around 1 μm, strong bands show obvious distortion in band shape. This occurs because at the wavelength of the absorption, the light refracted into the film is substantially reduced in intensity and the light reflected at the medium/film interface contributes an important proportion of the light arriving at the detector. The Kramers–Kronig relation may also be used for the analysis of the markedly distorted spectra of relatively thick films as well as for the estimation of film thickness.[43]

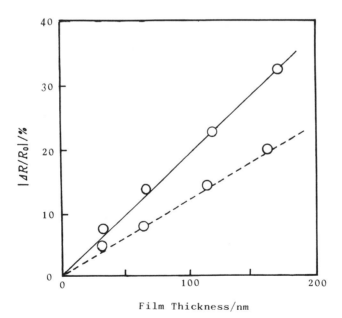

Figure 13. Relation between film thickness of copper oxides and band intensity. ——, CuO (560 cm^{-1}); − −, Cu$_2$O (645 cm^{-1}) (adapted from Ref. 17).

Figure 14 shows the C–O–C band in the reflection spectra of poly(2,6-dimethyl-1,4-phenylene oxide) films electrodeposited on copper.[49] In the spectrum of a film of thickness 1 μ, a shoulder appears at the low wavenumber side and increases in intensity with increasing thickness, as can be seen in Fig. 14c. This change may lead to an erroneous conclusion that a new C–O–C group is present in thick films and increases in relative quantity with film thickness. Reflection spectra of the polymer films of various thickness were calculated through the aid of the Kramers–Kronig transformation and the obtained spectra are shown in Fig. 15. This figure clearly indicates that the shoulder does not stem from the coexistence of two dissimilar C–O–C groups, but is just a case of distortion of band shape.

When a recorded spectrum of a relatively thick film shows very strong features whose reflectance is about 50% or less at the peak, care must be taken in interpretation because the features may be distorted in shape. The diagnosis

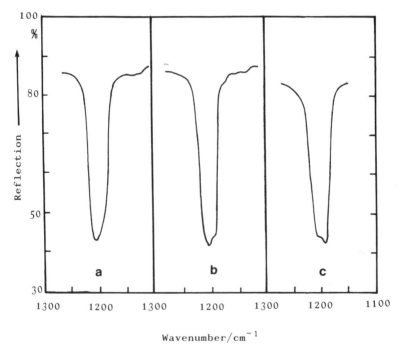

Figure 14. Observed infrared reflection spectra of poly(2,6-dimethyl-1,4-phenylene oxide) films on copper. Film thickness of (a) 0.4 μm, (b) 1.0 μm, and (c) 1.5 μm (adapted from Ref. 49).

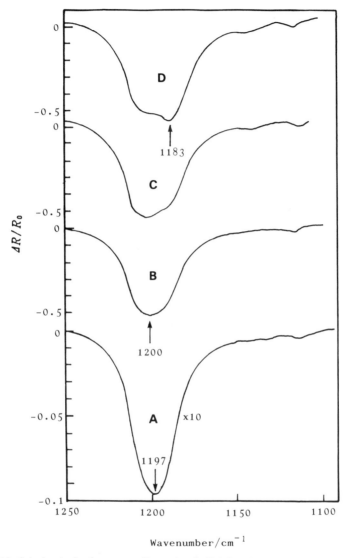

Figure 15. Calculated reflection spectra of isotropic poly (2,6-dimethyl-1,4-phenylene oxide) films of various thickness on copper. Film thickness of (A) 0.03 μm, (B) 0.3 μm, (C) 1.0 μm, and (D) 1.5 μm (adapted from Ref. 49).

of band distortion can be done conveniently by rotating the plane of polarization of the incident IR beam. If the feature is distorted, a remarkable change in band shape and a simultaneous shift of band peak take place upon rotation of the plane of polarization from parallel to perpendicular to the plane of incidence. When the feature is not distorted, the rotation of the plane results only in a decrease in intensity of the feature. Relatively undistorted spectra are often obtained from thick films on metal surfaces by the use of low incidence angles.

Reflection spectra of ultrathin films on highly reflective metals are obtained with highest sensitivity by the use of the optimum incident angle. However, if the film has a thickness exceeding the linear approximation limit, the highest sensitivity is obtained at an incident angle lower than the optimum angle. Figure 16 shows reflection spectra of a mercaptobenzothiazolato–Cu(I) film on copper recorded at various incident angles. Although the optimum angle of incidence for a thin film on copper is 89°, this figure shows that the incidence of IR light at 70° gives the strongest intensity of the absorption bands.

It would be informative here to show an example of marked change in a spectrum arising from molecular orientation in a surface film. Molecules in films on metal surfaces are generally oriented, unless the film is amorphous, because of the interaction of the metal surface with molecules in the first layer. Since the electric field oscillating at the metal surface is polarized normal to the surface, as mentioned in Section 1.1, reflection spectra of surface films of oriented molecules differ from the absorption spectra of the molecules in the isotropic state. A reflection spectrum of a thin maleic acid film on Incoloy (Ni–Fe–Cr–Mo alloy) is shown in Fig. 17.[50] An absorption spectrum of fine maleic acid crystals is also shown in this figure for the purposes of comparison. All the bands of the acid have nearly disappeared in the reflection spectrum except the starred bands at 870, 925, and 990 cm^{-1}. The band at 870 cm^{-1} is assigned to the C–H out-of-plane deformation vibration, and the other two bands arise from the O–H out-of-plane vibrations. Since only the modes with normal component of dynamic dipole moment can produce IR absorptions, in accordance with the surface selection rule, the obtained reflection spectra shows that maleic acid has a planar structure and is oriented with its plane parallel to the metal surface.

1.4. Applications

1.4.1. Real Surfaces

The observation of clean surfaces under UHV provides us with information indispensable for the basic understanding of surface phenomena. However,

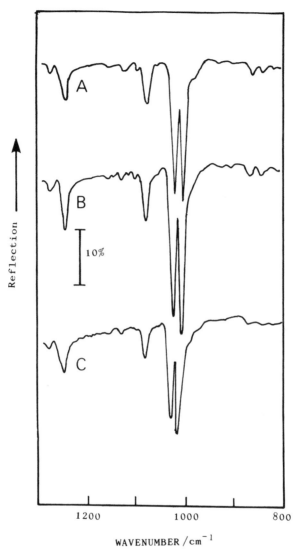

Figure 16. Reflection spectra of a mercaptobenzothiazolato–Cu(I) film (thickness of ~1 μm) on copper at various incident angles. Angle of incidence (A) 80°, (B) 70°, and (C) 55° [adapted from W. Suëtaka, ed., *Interface and Colloid,* Chemical Society of Japan, Maruzen, Tokyo (1977), p. 173.]

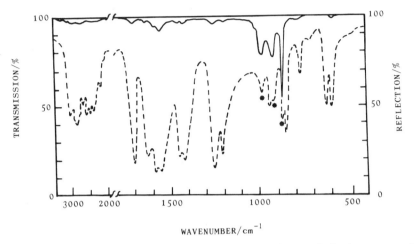

Figure 17. Infrared absorption and reflection spectra of maleic acid. ———, Reflection spectrum of a film on Incology; – –, absorption spectrum of powdered acid (Nujol and HCB mull) (adapted from Ref. 50).

the observation of clean surfaces has several important drawbacks. Metastable reaction intermediates adsorbed on metal surfaces, for example, are rarely observed in UHV at the elevated temperatures where the reaction proceeds. The observation of species physisorbed on clean surfaces requires refrigeration of the sample. The observation of real surfaces functioning in gaseous and liquid media is therefore equally important.

Infrared spectroscopy is widely used for observing samples in gaseous media, because as is well known, the gaseous homonuclear diatomic molecules and rare gases do not interfere with the observation of the surface IR spectrum, and signals of absorbing gas molecules can be eliminated by the use of double-beam spectrometers or the application of the polarization modulation method. Consequently, the IR reflection method is applied to the investigation of both real and working surfaces in gaseous media. Furthermore, when polarization and potential modulation methods are applied to the reflection measurement, signals of absorbing liquid media are reduced substantially. Hence, IR reflection spectroscopy is also used for the *in situ* observation of reactions on electrode surfaces, as will be illustrated in the following sections.

The IR reflection observation of thin surface films gives information about the species and their orientation in the films, thermal and photochemical changes in surface species, growth mechanism of the films, and film thickness.

This information is of great interest to technologically important areas such as corrosion, catalysis, lubrication, adhesion, Langmuir–Blodgett (LB) films, and growth of semiconductor films; significant research on real surfaces using IR reflection spectroscopy has been reported.

Langmuir–Blodgett films are important not only in technical applications such as sensors, microelectronics, optical devices, and spacers, but also in the basic study of film structures and phase transitions.[51-53] Infrared spectroscopy has been employed in the investigation of molecular orientation in LB films for many years.[54] Francis and Ellison investigated the orientation of metal stearates in LB films and the mutual interaction between carboxylate ions in the first IR reflection measurement.[4]

The orientation of aliphatic chains and functional groups in films has been determined by the parallel use of reflection and transmission IR measurements.[55,56] The attenuated total reflection (ATR) method is also used to determine the molecular orientation.[57] Besides the orientation of various functional groups, information about the molecular packing was obtained from the band splitting caused by intermolecular interactions.[55,57]

The molecular orientation can also be estimated from a comparison with the spectrum of an isotropic sample.[58] If the intensity of an absorption band in the isotropic sample is obtained or calculated, the angle between the surface normal and the direction of dynamic dipole moment θ can be estimated from the relation[58]

$$I_{obs}/3I_{calc} = \cos^2 \theta \qquad (17)$$

where I_{obs} and I_{calc} are the experimentally observed intensity from the anisotropic film and the calculated value for the isotropic sample, respectively. The photopolymerization of long chain compounds having unsaturated groups is used to obtain LB films of high mechanical strength. Equation (17) was used to determine the change in orientation arising from the photopolymerization of octadecyl methacrylate films, and it was deduced that, whereas the orientation of methacryl residue changes upon polymerization, the long chain of alkyl group remains nearly unchanged in orientation.[59] This is in agreement with the recent finding that the variation in the chain-terminating functional group has relatively little effect on the structure of the region of long hydrocarbon chains.[60]

In the Langmuir–Blodgett process, however, long chain compounds of particular structure should be used. The spontaneous adsorption technique can remove this difficulty,[61] and molecular assemblies with structures similar to those of LB films are obtained by this method.[58,62] Long chain alkanoic acids, n-alkyl sulfides, and disulfides are spontaneously adsorbed on metal surfaces from solution to form assemblies that are remarkably stable in air.[63] Spontaneous adsorption of n-alkyl anhydrides on oxide-covered aluminum was studied with the reflection IR method, and the formation of n-alkyl ester-type species was found in the surface film in addition to carboxylate species.[64]

Debe[65] has developed another approach for determining the molecular orientation from the intensities of reflection spectra and has applied this approach to the estimation of the orientation of the photomasked surfactant and photoconductors on aluminum.

The surface of metals exposed to air is covered with a layer composed of various compounds, such as corrosion products, conversion coating films, and protective films formed by the use of corrosion inhibitors. When the surface layer is thick, a quantity sufficient for preparing a KBr micropellet can be scraped off with a razor blade. Although well-resolved IR spectra are obtained with the KBr pellet technique, this technique cannot be applied to ultrathin films because of the low quantity of material irradiated. In addition, the film compounds may alter in a reaction with KBr. Infrared reflection spectroscopy has been employed in the measurement of oxides formed on metal surfaces. Poling[40] demonstrated that oxides grown on iron and copper could be discriminated by reflection spectroscopy. The first information about the composition of thin surface films of tin-free steels was obtained from the measurement of IR reflection spectra.[66] The oxidation of aluminum single crystals and stainless steels at various temperatures were studied with reflection IR spectroscopy.[67,68] Corrosion products on low alloy steel surfaces were studied by the reflection method, and the spectra obtained give information about the chemical history of the corrosion process.[69] The reflection spectra from inorganic films on metal surfaces were distorted, as discussed in the preceding section. Ottensen[70] discussed the distortion of band shape and frequency shift in the spectra of thin oxide films on metals and gave hints for the quantitative determination of film thickness.

Several organic compounds are used as inhibitors of metallic corrosion. These compounds protect metal surfaces from corrosion through adsorption or protective layer formation on the metal surfaces. Since these protective layers are generally composed of organic or organometallic polymers, IR reflection spectra can be obtained without noticeable distortion from the layers. Benzotriazole is an effective corrosion inhibitor of copper and its alloys, and reflection IR spectroscopy has been employed in the investigation of the mechanism of corrosion inhibition by this material and its related compounds.[71–74] Mercaptobenzothiazole is another effective corrosion inhibitor for copper. Infrared reflection spectra have shown the presence of thin polymeric films of mercaptobenzothiazolato–Cu(I) complex on the copper surfaces treated with the inhibitor.[75] This film is water-insoluble and acts as a barrier against corrosive environments. Acetylenic alcohols control the dissolution of iron through the formation of a barrier layer of hydrocarbon polymer which is bearing COOFe and COFe groups.[76] Thibault and Talbot have deduced from reflection IR measurements that the protective action of sulfoxides against corrosion of iron in acidic solutions is due to the formation of protective films of organic sulfides.[77]

Addition of surfactants to mineral oil reduces remarkably the coefficient of friction in boundary lubrication by the oil. Since the behavior of molecules in a thin film formed between a metal plate and a solid that is transparent in IR region can be observed *in situ* under pressure by the IR reflection method, the orientation of *n*-octadecane in the film was observed by the reflection method at various temperatures with or without long chain fatty acid addition.[78]

The results obtained from a film of *n*-octadecane between an electrolytic iron and a KBr plate are shown in Fig. 18. In this figure the intensities of symmetric and antisymmetric stretching vibrations of the CH_2 groups of n-octadecane are shown for *p*- and *s*-polarized light. If the film is isotropic, the

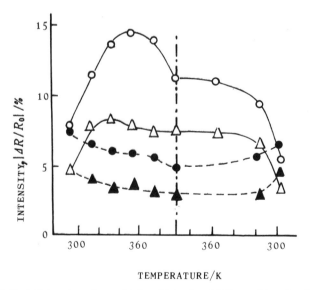

Figure 18. Temperature dependence of infrared intensity of CH_2 stretching vibrations of *n*-octadecane in a thin film. Film thickness = 40 nm, applied pressure = 490 kPa. ——, Measured with *p*-polarized light; – –, measured with *s*-polarized light; O, ●, antisymmetric vibration band; Δ, ▲, symmetric stretching vibration band (adapted from Ref. 78).

absorption intensity should be much stronger in the measurement with p-polarized light than in the measurement with s-polarized light. At room temperature, however, the absorption intensity remains nearly constant, independent of the polarization of the incident light. This means the hydrocarbon film is not isotropic under the applied pressure, but that the CH_2 groups are oriented nearly parallel to the surface and consequently the carbon chain is normal to the iron surface. Upon increasing the temperature, the absorption intensity of p-polarized light increases rapidly and that of s-polarized light decreases, indicating the dissipation of the orientation.

The results obtained from a film containing stearic acid are shown in Fig. 19. This figure shows that the addition of the acid promotes the orientation of hydrocarbon molecules and suppresses the disordering of the molecules when friction brings about an increase in temperature. The use of gold in place of iron resulted in the loss of the effect of fatty acid addition. A decrease in chain length of the fatty acid also attenuates the effect of fatty acid addition. All of these observations suggest that the fatty acid added to the hydrocarbon film

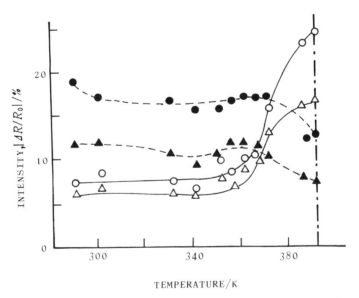

Figure 19. Change in intensity with temperature of CH_2 stretching bands of an n-octadecane film containing stearic acid (18%). Film thickness = 60 nm, applied pressure = 490 kPa. ——, p-polarized light; − −, s-polarized light; O, ●, antisymmetric vibration band; Δ, ▲, symmetric vibration band (adapted from Ref. 78).

chemisorbs on the metal surface and makes the orientation of hydrocarbon molecules more perfect, maintaining it even at elevated temperatures.

Extreme pressure agents reduce friction under high load and high speed. The working mechanism of the agents is not sufficiently clear, but they probably produce a layer of compounds susceptible to shear fracture on the metal surface at high temperatures generated under the severe friction, reducing the coefficient of friction. Francis and Ellison investigated the surfaces of silver and steel with IR reflection spectroscopy after immersing the metals in hydrocarbon solutions of zinc dialkyl dithiophosphate, an effective extreme pressure agent.[79] Zinc oxide, sulfate, and adsorbed molecules of the zinc additive were detected in the spectra obtained. The compounds on the parts of high compression engines formed by the addition of an extreme pressure agent was investigated through FTIR reflection spectroscopy.[80] Reaction products derived from the agent on the surfaces changed in relative quantity depending upon the operation mode of the engine. The reaction of carbon tetrachloride with atomically clean Fe(110) surfaces has been investigated in relation to extreme pressure lubrication, and the formation of the lubricant, $FeCl_2$, is postulated on defect sites.[81] The behavior of extreme pressure agents was also investigated by means of infrared emission spectroscopy and will be mentioned in Chapter 4.

The chemical bond formed between adhesive molecules and oxide layers on metal surfaces is an essential problem in adhesion. Boerio and Chen observed the reflection spectra of epoxy resin thin films of about 1.5 nm thickness on iron and copper mirrors, and have deduced that the epoxy resin molecules are adsorbed with a vertical conformation and a bond is formed between a single oxirane oxygen atom and the surface.[82] They also studied the adsorption of dodecanoic acid and 1-octadecanol on the surfaces of oxidized metals in connection with adhesion and lubrication.[83] The spectra obtained showed the adsorption of the fatty acid through the formation of the carboxylate group and its orientation perpendicular to the substrate surface. Hydrogen bond formation between carbonyl groups of cyanoacrylate molecules and O–H groups at the surface of the oxide layer on aluminum was deduced from the shift of absorption bands arising from the decrease in thickness of the adhesive film.[84]

When adhesion is good, the adhesive failure occurs in the cohesive mode. High mechanical strength of the adhesive layer, therefore, is important for obtaining high adhesive strength. The shear adhesion strength of a joint made of aluminum coupons with cyanoacrylate changed as a function of surface treatment. The shear strength was lowest on an electrolytically polished surface and increased with an increase in pore density of the oxide layer formed by anodic oxidation. In addition, whereas the fractured surface of the adhesive showed the quasi-cleaved fracture on the electro-polished aluminum, dimple patterns were observed on the aluminum covered with an oxide layer of high pore density. Reflection IR spectra of adhesive thin films on aluminum coupons showed that the inclination of the C=O bond of cyanoacrylate to the metal surface increased, parallel to the decrease in adhesive strength.[84] From these facts, it was deduced that the hydrogen bond formed at the pore wall was responsible for the molecular orientation. The preferred molecular orientation in the adhesive layer may correspond to the increase in mechanical strength of the polymer layer.

Xanthates are widely used in flotation, and the adsorption of xanthates from aqueous solutions on metal surfaces has been studied by reflection IR spectroscopy.[6,85] At the lowest submonolayer coverage, the adsorbed xanthates are randomly oriented. The degree of orientation changes depending on the thickness of the adsorption layer, and finally multilayer films of randomly oriented Cu(I) xanthate are formed.[85]

Highly oriented films are formed by vapor deposition in addition to Langmuir–Blodgett and spontaneous adsorption techniques.[86–89] Debe et al.[90] employed IR reflection spectroscopy in the investigation of orientation and crystallinity of copper phthalocyanine (CuPc) films formed by vacuum sublimation and physical vapor transport. The results obtained from films deposited from the vapor phase suggested that natural self-ordering took place on amorphous substrates. The orientation, however, changed completely when the film was grown on a previously oriented seed film. The isoepitaxy of CuPc on metal-free phthalocyanine and the homoeepitaxy on CuPc were demonstrated. Vapor-deposited thin films of p-chlorophenylurea were investigated with x ray diffraction and reflection IR spectroscopy.[91] A highly preferential orientation of this compound with {001} planes parallel to the substrate surface was observed. The orientation is better in the thinner films than in the thicker films.

Reflection IR spectroscopy was also used in the study of the effect of gravity on the orientation and crystallinity of CuPc thin films.[92] Besides IR reflection spectroscopy, grazing incidence x ray diffraction and visible-near IR reflection spectroscopy were employed in the investigation of thin films, and comparisons were made between the sample formed by a physical vapor transport method in the Space Shuttle Orbiter and the sample obtained in the laboratory on the ground. The results show that, whereas the microgravity-grown films contained predominantly a new polymorphic form of CuPc, the ground controls contained mixtures of α and β polymorphs. Differences in molecule orientation were found between the space-grown films and the laboratory-grown samples.

Infrared reflection spectroscopy is also a potential tool for understanding the growth mechanism of semiconductor films. Hydrogenated amorphous silicone (a–Si:H) films photochemically growing on aluminum substrates were investigated in a flow reactor with reflection spectroscopy.[93] In the initial stage of deposition, IR bands assignable to SiH_3 and SiH_2 were detected and persisted without variation of intensity during further deposition. The bands due to SiH groups appeared later and continued to increase in intensity. The results suggest the existence of a hydrogen-rich layer. The growth of a–Si:H in an rf glow-discharge plasma was studied by the real-time *in situ* observation of IR reflection spectra.[93] In this work, the deposition was begun with the introduction of SiD_4 instead of SiH_4 into the reactor. The deposition was continued using SiH_4, after evacuation of gaseous SiD_4. The use of the deuterated compound resulted in a remarkable decrease in the intensity of the SiH_3 and SiH_2 bands. Analyses of this and the other results have led to a deduction that the hydrogen-rich species are present in the Si:H/Al interface in addition to the outermost layer of the growing semiconductor film.

The surface films formed on GaAs substrates in metallorganic vapor phase epitaxy were observed with reflection IR spectroscopy. The measurement was made at 300K in a background trimethylgallium atmosphere at 10^{-4} Pa. Three positive features were observed at 1050, 1174, and 1420 cm^{-1}. The latter two features may be assigned to methyl groups of surface species, but physisorbed trimethylgallium may also exhibit these features.[94] This work shows that reflection IR spectroscopy may be instrumental in understanding the growth mechanism of compound semiconductors.

1.4.2. Clean Surfaces

Techniques using electrons and ions as probes are widely employed in the investigation of clean and well defined surfaces in UHV. Infrared spectroscopy has the advantages of high resolution and of non-charging of samples compared to the electronic and ionic techniques. However, it is lower in sensitivity than the latter two techniques, because photons interact weakly with matter. Hence in the early IR investigations of adsorption on metal surfaces, polycrystalline or recrystallized ribbons and evaporated metal films were used as adsorbent and very strongly absorbing CO or carboxylate ions as adsorbate.[95–104] For example, Yates et al.[96,97] studied adsorption of CO on an atomically clean polycrystalline tungsten ribbon with IR reflection spectroscopy and thermal desorption measurement. From the behavior of the band due to the weakly bound α–CO state, they found that this state can be resolved into two states, both of which involve sp-hybrid carbon bonded to the tungsten surface.

The efforts of spectroscopists have progressively improved the sensitivity through the use of improved detectors, Fourier transform spectrometers, and various modulation techniques. The adsorption of a variety of species on single crystal surfaces can now be observed with IR reflection spectroscopy, and numerous works have been reported. This section deals with selected examples obtained without applying the modulation techniques. Detailed theoretical discussion is outside the scope of this volume.

Nguyen and Sheppard[105] reviewed the CO adsorption on dispersed metal and showed that the stretching frequency of adsorbed CO can be classified into four groups corresponding to the adsorption sites: 2000–2130 cm^{-1} (terminal CO), 1880–2000 cm^{-1} (two-fold bridged CO), 1800–1880 cm^{-1} (three-fold bridged CO), and <1800 cm^{-1} (four-fold bridged CO). The adsorption sites of CO on extended metal surfaces can be discerned by the use of this classification.

Krebs and Lüth studied the adsorption of CO on Pt{111} with a double-beam spectrometer equipped with a cooled PbSnTe detector and observed for the first time in IR reflection spectroscopy two absorption bands near 2100 cm^{-1} and 1870 cm^{-1} assignable, respectively, to CO adsorbed at on-top and bridge sites.[106] Campuzano and Greenler studied the adsorption of CO on Ni{111} by means of IR reflection spectroscopy combined with LEED and the other methods. The stretching frequency of adsorbed CO appears at 1817 cm^{-1} at low CO coverage ($\theta = 0.05$). The absorption peak shifts to higher frequencies with increased coverage, as shown in Fig. 20. This figure shows that when the coverage is lower than ~0.2, the absorption peak is located at a frequency lower than 1850 cm^{-1}, suggesting that CO is bound to three atoms of nickel. At $\theta \sim 0.25$, the peak frequency abruptly moves to about 1900 cm^{-1}, corresponding to two-fold bridge site adsorption. At $\theta = 0.57$, an additional absorption band (not shown in this figure) appears at 2045 cm^{-1} arising from terminally adsorbed CO. The LEED observation showed a c(4×2) structure at $\theta = 0.50$ and a ($\sqrt{7}/2 \times \sqrt{7}/2$)R19.1° pattern at $\theta = 0.57$. The solid line in this figure shows the deduced frequency shift due to dipole-dipole interactions, which is rather small.[107] A similar frequency jump was observed on a Pd{111} surface with the polarization modulation method.[108] The surface-site-specific absorption and frequency shift with coverage are reviewed by Hoffmann.[35]

More recently, the CO–Ni{111} system was carefully investigated by IR reflection spectroscopy combined with LEED and the other surface-sensitive techniques. At the lowest coverage ($\theta = 0.015$), two bands appear showing the existence of two-fold coordinated CO species in addition to the three-fold coordinated CO species. The frequency shift observed at 90 K is shown in Fig. 21 against CO coverage. At high coverage ($\theta \cong 0.57$), a double CO band was

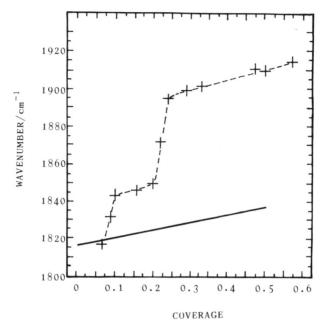

Figure 20. Frequency of infrared band peak of CO as a function of coverage on a Ni{111} surface [Reprinted from J. C. Campuzano and R. G. Greenler, *Surf. Sci.* **83**, 301 (1979)].

seen after annealing at 240 K in addition to the terminal CO band, implying that a small part of the surface was covered with c(4 × 2) structure domains.[109]

The above-mentioned sites of CO adsorption on Ni{111} surfaces are interchangeable.[110] Reflection IR and work function measurements show that, while more stable bridge-bonded CO decreases in quantity with increasing temperature, terminally bonded CO increases in surface coverage within the temperature range of 90 to 288 K. Figure 22 shows the corresponding spectral change. No terminal band appears at $\theta = 0.50$ in the observed temperature range, in agreement with the above-mentioned work.[109] The binding energy difference between the two kinds of CO and the zero-coverage dipole moment of both the species were determined. Whereas a strongly nonlinear intensity increase is observed for the IR band of bridged CO with coverage above $\theta \simeq 0.2$, terminal CO exhibits a more linear intensity behavior with its coverage.

The structure of the high-coverage CO adlayer on Pt{111} and Pd{111}

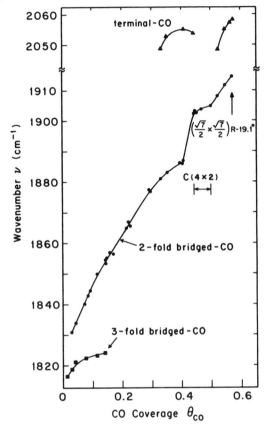

Figure 21. Observed frequency of all three bonding modes of CO on Ni{111} as a function of coverage. All spectra were recorded at 90 K following annealing at 240 K [Reprinted from L. Surnev, Z. Xu, and J. T. Yates, Jr., *Surf. Sci.* **201**, 1 (1988)].

was studied by IR reflection and LEED measurements, and the formation of a coincident site lattice in which the molecules are adsorbed on high symmetry sites has been deduced.[111]

Hydrogen adsorption on W{100} and Mo{100} surfaces induces complicated substrate reconstructions.[10] It has also been found that hydrogen adsorbs exclusively on two-fold bridge sites on W{100} and Mo{100} surfaces.[112,113] The surface structure formed by hydrogen adsorbed on Mo{100} changes depending upon the temperature and coverage, correlating with the

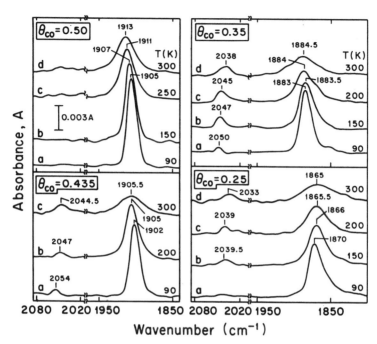

Figure 22. The reversible temperature dependence of the IR spectra of CO adsorbed on Ni{111} [Reprinted from L. Surnev, Z. Xu, and J. T. Yates, Jr., *Surf. Sci.* **201**, 14 (1988)].

reconstructed geometry of the substrate.[113] The first overtone of the wagging vibration of hydrogen adsorbed on W{100} and Mo{100} shows an intensity stronger than that theoretically expected and has a highly asymmetric line shape.[114,115] A coupling between the hydrogen vibration and the continuum absorption due to surface electronic transitions was concluded from the results obtained.[115]

The adsorption site of CO on the reconstructed (hex) Pt{100} surface is different from on the unreconstructed Pt{100} surface, and initially only the linear CO band appears at a wavenumber higher than on the unreconstructed surface. A new linear CO band appears on the reconstructed surface with increasing exposure to CO and dominates the spectrum. Isotopic dilution measurements indicate that the appearance of the second band would correspond to the creation of (1 × 1) islands as a result of the partial removal of the reconstructed structure.[116] The reconstructed Pt{100} surface is immediately lifted upon adsorption of NO above 210 K. Below 200 K, the reconstructed structure remains unchanged until a critical NO coverage is reached, which is dependent on temperature.[117]

Electronegative and positive atoms present on metal surfaces exert pronounced effects on the adsorption state of CO. The investigation of these effects is important in understanding the role of poison and promoter in heterogeneous catalytic reactions involving CO. Long-range interactions such as chemical and electrostatic effects are proposed between CO and the adsorbed atoms. For example, the electron transfer from adsorbed electropositive atoms to the surface is considered in the chemical effect. The electron transfer makes more electrons available to be donated to the anti-bonding orbitals of CO, resulting in the loosening of the C–O bond and the simultaneous strengthening of the M–C bond.[118] In the electrostatic effect, the attraction of electrons by a CO molecule from the surface is taken into account. The attraction of electrons produces a dipole which interacts attractively with the electropositive atom-induced dipole.[119] Short-range interactions causing surface complex formation may occur between the adsorbed electropositive atom and the CO molecule, and formates and carbonates have recently been detected, as described below.[120]

Yates *et al.* studied the adsorption of CO on a sulfur-covered Ni{111} surface. Adsorbed CO exhibited an infrared band at a frequency higher than on the clean Ni{111} surface. This and other data were interpreted as indicating a local short-range interaction between CO and S on the Ni surface.[121] The presence of slightly larger range S···CO interaction was also deduced.[122]

Both long-range and short-range interactions were observed between potassium and CO on the Ni{111} surface. It is estimated that the long-range K···CO interaction, which is electrostatic in nature, extends to ~25 CO molecules from one potassium atom. The short range K···CO interaction due to the formation of a K–CO complex was suggested, although strong electrostatic effects may provide a better description of the interaction.[123,124]

A Xe atom adsorbed on a metal surface interacts with coadsorbed CO. Figure 23 shows the shift of the absorption band of CO adsorbed on Ru{0001} resulted from the adsorption and desorption of Xe. The observed shift of 38 cm^{-1} is explained by electrostatic interaction.[125] Additional studies of the Xe–CO interaction have been carried out on Ni{111}, reaching similar conclusions.[126]

The lateral interaction between adsorbed CO molecules gives rise to the shift of absorption peak as well as the change in band shape. This topic is reviewed in detail by Hoffmann.[35] Examples of the peak shift due to the lateral interaction can be seen in Figs. 20 and 21. The line broadening is divided into inhomogeneous and homogeneous broadenings. Inhomogeneous broadening

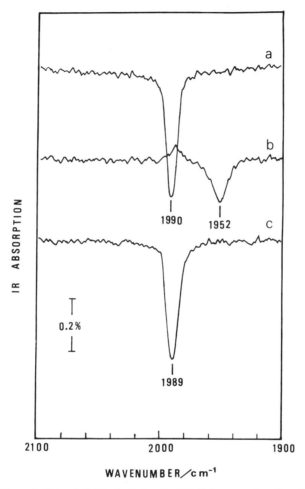

Figure 23. Change in IR band of CO adsorbed on Ru{0001} due to coadsorption of Xe at 75 K. (a) Spectrum of adsorbed CO at a coverage θ = 0.03, (b) after subsequent Xe adsorption, and (c) after Xe desorption and recooling to 75 K. (Resolution = 2 cm^{-1}) [Reprinted from F. M. Hoffmann, N. D. Lang, and J. K. Nørskov, *Surf. Sci.* **226**, L48 (1990)].

comes from the variety of vibrational frequencies of molecules adsorbed at various adsorption sites and of intermolecular distance between adsorbed molecules. Homogeneous broadening stems from the damping of vibration as a result of interaction of the vibrating adsorbate with the metal surface through phonon coupling and electron–hole pair creation.

The temperature dependence of the vibrational band shape for CO adsorbed at both the two-fold bridge and on the top sites of a Ni{111} surface was studied over the temperature range of 80 to 300 K. Both the bands remained symmetric and unchanged in position independent of temperature, suggesting that they are homogeneously broadened. These bands showed opposite behavior to one another in temperature-dependent broadening. This behavior is explained according to a vibration dephasing model.[127] Higher quality reflection IR spectra were obtained by Persson and Ryberg and were interpreted as being due to vibrational dephasing.[128]

The infrared investigation of multilayer adsorption on metal surfaces is of importance with respect to catalytic reaction and interlayer interaction. In catalytic reactions, multilayers of adsorbates are sometimes formed on the catalyst surface. Infrared spectroscopy is capable of detecting small changes in the vibration of adsorbates and is appropriate for examining the weak interlayer interaction.

Infrared reflection spectra have shown that methanol is chemisorbed on Pt{111} surfaces by surface bonding at the oxygen end at 90 K. Hydrogen bond formation was detected for the layer physisorbed on the chemisorbed first layer. Outerlayers condensed on the second layer are amorphous but show a phase transition into the crystalline α-ice modification of methanol upon annealing to 125 K.[129]

Two types of interaction are detected between an NH_3 overlayer and an underlying CO layer on Ni{111}. NH_3 molecules are adsorbed on the saturated CO layer in forming strong NH–OC bonding. The formation of an NH_3 overlayer brings about a CO site interconversion from terminally bound CO to bridge-bound CO species in the CO underlayer. Figure 24 shows the change in intensity of IR bands of $^{13}C^{18}O$ adsorbed on Ni{111} upon introduction of overlayer NH_3. The decrease in intensity of the 1963 cm^{-1} band (terminally bound CO) is accompanied by the intensity increase of the 1833 cm^{-1} band (bridged–CO). The increase in overlayer coverage of NH_3 from 0.15 to 0.32, which produces the intensity increase in the NH_3 bands at 1644 and 1466 cm^{-1}, results in a further decrease in intensity of the terminal CO band. It is deduced from the results obtained that the polar top layer imposes a strong positive-outward electric field on the underlayer, which causes a rehybridization of chemisorbed CO species and results in the CO site interconversion. This phenomenon is not specific to NH_3, but is related to the electric field induced by the physisorbed second layer, as reflected in the work function decrease.[130,131]

The characterization of finely divided metal surfaces is important with respect to the reaction on a dispersed metal catalyst, and among surface sensitive techniques, infrared spectroscopy can be applied directly not only to the study of dispersed metal catalysts but also to the investigation of clean and well

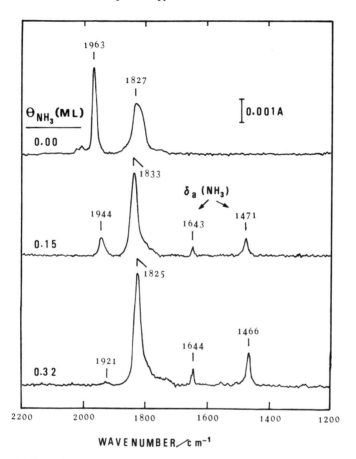

Figure 24. Change in intensity of IR band of CO adsorbed on Ni{111} depending upon NH_3 overlayer coverage T_{ads} = 90 K, T_{ann} = 150 K [Reprinted from Z. Xu, J. T. Yates, Jr., L. C. Wang, and H. J. Kreuzer, *J. Chem. Phys.* **96**, 1628 (1992)].

defined metal surfaces. The infrared frequency of CO adsorbed on supported Pd can therefore be utilized for the determination of particular low index planes in comparison with the results obtained from Pd single crystal surfaces.[132]

Infrared spectra from CO adsorbed on Pd/Al_2O_3 were compared with adsorption on the Pd{111} surface, and it was found that CO molecules are adsorbed initially on {111} facets, because there is a close resemblance between the spectra of supported and single crystal Pd. However, when the surface of supported Pd is precovered with ethylidyne, the IR features of adsorbed CO lose the above-mentioned resemblance. A model is proposed in which ethylidyne blocks all sites for CO adsorption on {111} facets on Pd crystallites.[133]

Finely divided metals present edge and corner sites in relatively high density. The observation of CO adsorbed at steps and kinks of clean metal surfaces should provide information valuable for the basic understanding of adsorption on finely divided metal particles. The adsorption of CO on a Pt{533} crystal surface, a surface with 4-atom wide {111} terraces and {100} steps, was investigated with IR reflection spectroscopy at 85 K. Initially, CO adsorption takes place at the step sites, and then on the {111} terraces with a reduced sticking coefficient. No bridge site adsorption was observed on the stepped surface. The frequency of the step CO singleton is close to that of the terrace CO singleton, but the IR extinction coefficient for CO adsorbed at the step site is a factor of 2.7 greater than for CO in the two-dimensional domains at high coverage.[134] The strength of the electric field near a step in a metal surface has been calculated, and an enhancement in the electric field intensity ranging from 2.0 to 2.4 is obtained for CO adsorbed at the step site. This enhancement of the electric field could give rise to the increase in observed IR intensity of molecules adsorbed on the step.[135] Recently, bridged CO on the steps of Pt–s[4{111} × {100}] was observed.[136]

The CO adsorption on Pt{111}, {533}, and {432} surfaces has been compared with the adsorption on platinum particles. The bands appearing at 2081, 2070, and 2063 cm^{-1} are assigned to CO terminally bound to face, corner, and edge atoms of metal particles, respectively.[137] With respect to the adsorption on the stepped surface, the wavenumber of the isolated molecules has been deduced as 2074, 2065, and 2060 cm^{-1} for CO adsorbed at terrace, step, and kink sites, respectively.[138]

CO adsorbed on the stepped Pt surfaces exhibits two bands at intermediate coverage. A lower frequency band persists at low coverage and a higher frequency band at high coverage. The behavior of the two bands was compared with the calculation based on harmonic oscillator models, and it has been shown that the experimental data can be explained by a coupled-dipole model in which the effects of electronic polarizability, the tilted orientation of CO molecules at step sites, and the electric field enhancement at step sites are taken into account.[138,139] A model for calculating normal modes and infrared intensities was formulated for CO adsorbed on extended crystal surfaces and small particles of platinum. The adsorbed molecules are treated as coupled harmonic oscillators and the coordination number of the platinum atoms, on which CO adsorbs, is taken into account. The model is applied both to CO adsorbed on single-crystal surfaces of platinum and to CO on small spherical particles of platinum.[140]

Infrared reflection spectroscopy is used for obtaining information about the mechanism of reactions proceeding on bare and oxide-covered metal surfaces. The oxide-covered metal is used as a model system for investigating reactions on metal oxide surfaces because of the ease of obtaining clean surfaces. In the observation of surface reactions, a small chamber, which is combined with a large chamber for preparation and surface characterization, is generally used for IR measurement at relatively high pressures. An example is shown in Fig. 25.[121,122] Similar configurations were used by Banholzer and Masel;[141] Hoffmann, Robbins, and Weisel;[142,143] Erley;[144] and Campbell and Goodman.[145] The signals from gas-phase molecules can be eliminated by the double-beam operation of the spectrometer (including the subtraction of the background spectrum separately recorded) or the polarization modulation method. The measurement of surface reaction is often performed at elevated temperatures, where emissions from molecules on the metal surface have an appreciable intensity in the long wavelength region. The emission is superimposed on the absorption, resulting in the distortion of the absorption band. The emission from the sample at moderately elevated temperatures, as well as the background emission from the IR cell, can be eliminated for the most part by the application of polarization modulation or by intensity modulation with a mechanical chopper placed between the sample and the infrared source.

Shigeishi and King investigated the oxidation of CO on a recrystallized {111} oriented Pt ribbon with IR reflection spectroscopy.[146] When the Pt surface was saturated with CO, the oxidation reaction did not proceed, even when oxygen was introduced over the Pt surface. The partial removal of adsorbed CO was necessary for the initiation of the reaction. This shows that oxygen must be chemisorbed for the reaction to begin and thus the reaction must proceed by a Langmuir–Hinshelwood mechanism. However, the introduction of CO on an oxygen-covered Pt surface caused immediate consumption of adsorbed oxygen, even though the surface was saturated with oxygen.

A "surface explosion" was observed in the reaction between NO and CO on Pt surfaces at 410 K.[147] The adsorption of NO, CO, and NO/CO mixtures on a Pt{100} surface was studied with IR reflection spectroscopy for elucidating this "surface explosion."[141] The results suggest that, when adsorbed CO has few NO neighbors, the reactivity of CO increases and the enhanced reactivity gives rise to the autocatalytic behavior of surface species, which could be responsible for the "surface explosion." The surface explosion may also be explained by a structural autocatalysis model, which involves the formation of a reactive intermediate structure involving Pt atom displacement.[147] In other studies, oscillations have been observed in the reaction between CO and NO over a Pt{100} surface. The oscillations are accompanied by the periodic lifting

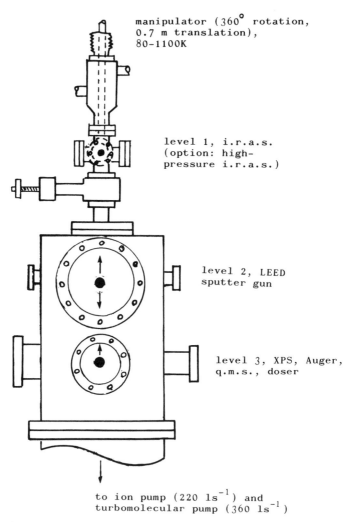

manipulator (360° rotation,
0.7 m translation),
80-1100K

level 1, i.r.a.s.
(option: high-
pressure i.r.a.s.)

level 2, LEED
sputter gun

level 3, XPS, Auger,
q.m.s., doser

to ion pump (220 ls^{-1}) and
turbomolecular pump (360 ls^{-1})

Figure 25. Ultrahigh vacuum chamber (applicable for the measurement at high pressures) for infrared reflection spectroscopy. The IR cell at level 1 can be separately pressurized with a gate valve [Reprinted from J. T. Yates, Jr., M. Trenary, K. J. Uram, H. Metiu, F. Bozso, R. M. Martin, C. Hanrahan, J. Arias, *Phil. Trans. R. Soc. London* **A318**, Fig. 3, p. 104 (1986), The Royal Society].

of the hexagonal surface reconstruction.[148] The interaction between CO and NO on Pt{100} and the correlation between the interaction and the lifting of the reconstruction was studied by IR reflection spectroscopy.[149]

The methanation reaction of CO on Ru{0001} was investigated through *in situ* observation of IR reflection spectra at elevated pressures (1.3×10^2–1.3×10^4 Pa) and temperatures (300–600 K).[150a] Vibrational spectra were recorded at various stages of the reaction, and the obtained features of adsorbed CO imply the presence of islands of surface carbon. The surface carbon at high coverages could suppress the adsorption of CO, resulting in the deactivation of the surface. The presence of hydrogen accelerates tremendously the dissociation of CO. This effect is probably attributable to the formation of water, which prevents the recombination of dissociated CO. Weakly adsorbed species also may exert significant effects on the reactivity of coadsorbates at elevated pressures. Mims *et al.*[150b] have observed that, at elevated pressures, a CO species weakly adsorbed in threefold hollow sites on Ru{0001} hinders the decomposition of ethylidyne species adsorbed in a neighboring surface site.

The state of CO adsorbed on metal surfaces is affected by the presence of adsorbed potassium, as mentioned above. In the hydrogenation reaction of CO on metal catalysts, the addition of alkali metals promotes the selectivity to higher members of hydrocarbons. Infrared reflection measurement is employed for *in situ* observation of the hydrogenation reaction of CO on a potassium modified Ru{0001} surface at elevated pressures.[120a,b] The obtained spectra revealed the formation of formate and carbonate intermediates bound with potassium during the reaction. This indicates that potassium not only promotes the dissociation of the C–O bond but also participates in the formation of reaction intermediates.[120b]

The isotopic mixing reaction between $^{13}C^{16}O$ and $^{12}C^{18}O$ on Ni{111} with preadsorbed potassium was investigated with IR reflection spectroscopy and TDS (Thermal Desorption Spectroscopy).[151] The results imply that potassium causes the dissociation of CO on Ni{111}, which is reversible at elevated temperatures, and the recombinant CO desorbs above ~450 K. Because the isotopic mixing reaction could not be detected in the observation of reflection spectra, the following sequential processes are likely to take place:

(a) dissociation of CO at Ni sites activated by K atoms.
(b) diffusion of C and O atoms away from the K-activated sites.
(c) recombination of C and O atoms followed by desorption.
(d) diffusion of CO from outside to refill the K-activated site.

In conclusion, the alkali promoter sites act as localized feeder centers for CO dissociation.

The decomposition of ethanol on Ni{111} was investigated in UHV using IR reflection, XPS, and TPD measurements. Ethanol, chemisorbed with the O–H bond nearly parallel to the metal surface, decomposes via a first step where O–H bond scission occurs. The ethoxy thus formed does not combine with chemisorbed hydrogen on Ni{111}, in contrast to the recombination of methoxy and hydrogen on a Ni{111} surface. The desorption of ethanol is accompanied by decomposition to the chemisorbed ethoxy species. The orientation of adsorbed ethanol before and after decomposition is discussed.[152]

The addition of an alkali salt to NiO promotes the oxidative coupling of methane to C_2 hydrocarbon species. For the fundamental understanding of alkali-promoted activation of methane on NiO, a model catalyst of K/NiO/Ni{100} was investigated with various surface-sensitive techniques. In the IR measurement, CO was used as a probing molecule and the presence of Ni^{3+} sites was suggested. The investigation of surface reactivity has shown that, while the reactivity of NiO toward H_2 is reduced as a result of the presence of potassium, its presence causes the reactivity with CH_4 to be enhanced.[153]

Time-resolved IR reflection spectroscopy should provide valuable information for elucidating the kinetics of heterogeneous catalytic reactions.[154] Time-resolved measurement is done with either a dispersive spectrometer or an interferometer. Dispersive spectrometers are used to monitor the change in sample reflectance as a function of time at a fixed frequency. This method is useful when the peak frequency does not shift with temperature and coverage. Benziger et al. investigated isosteric phase transformation of CO on Ni{100} with temperature programmed IR reflection measurement using a polarization modulation spectrometer. In this work the surface phenomena were followed with some-milliseconds time resolution.[155] When the peak frequency shift occurs during surface events, the use of a HgCdTe or InSb array detector would be appropriate.

Employing fast scanning FTIR spectrometers, time-resolved measurement can be performed in a total measurement time of 10–30 s by adding a small number of repetitive scans. When a fast single scan mode is used, a time-resolution of the order of 0.1 s can be obtained.[154,156]

Burrows studied the oscillating oxidation of CO over a Pt foil catalyst with time-resolved IR reflection spectroscopy under ambient pressures. The measurement was able to show a slow periodic variation in the number of active sites on the Pt surface, which is concurrent with the variation in reaction rate.[154] Chenery et al. investigated the desorption of CO from Cu{111} and the deprotonation of methanol on oxidized Cu{111} with time-resolved IR spectroscopy.[157]

Time-resolved IR measurements were utilized to investigate the kinetics of the dissociation of CO on Ru{0001} under methanation reaction conditions, and the reaction rate as well as the activation energy for the dissociation reaction was determined.[142] Time-resolved measurement of the reaction of CO with Ru{0001} indicates that the reaction is a non-steady-state process and results in CO disproportionation ($2CO \rightarrow C + CO_2$). The rate-determining step in the overall reaction is found to be the dissociation of the CO molecule, and the chemisorbed state of CO is a precursor to the dissociation.[158]

Time-resolved IR reflection observation of the CO oxidation reaction on a Ru{0001} surface has implied that, whereas the reaction proceeds by an Eley–Rideal mechanism involving reaction between gas-phase or weakly adsorbed CO and the O–(1 × 1)–Ru{0001} surface under oxidizing conditions (CO/O_2 ratio < 1), the reaction proceeds via a Langmuir–Hinshelwood mechanism between chemisorbed CO and oxygen under reducing conditions (CO/O_2 ratio > 2).[159] Figure 26 shows time-resolved IR spectra of Ru{0001} with preadsorbed oxygen ($\theta_0 = 0.6$) obtained upon exposure to 333 Pa of CO at 400 K. In the initial stage, the predominant peak of adsorbed CO appears at 2080

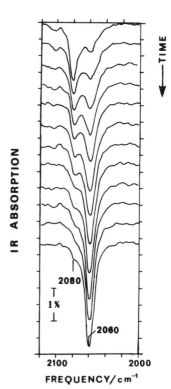

Figure 26. Time-resolved IR spectra of CO adsorbed on Ru{0001} with preadsorbed oxygen ($\theta_0 = 0.60$). The shift of CO band shows the removal of oxygen. The time interval between the spectra is 22 s. Temperature = 400 K, CO pressure = 332.5 Pa [Reprinted from F. M. Hoffmann, *J. Chem. Phys.* **90**, 2816 (1989)].

cm^{-1}, showing that most of the CO is coadsorbed with oxygen. Its intensity decreases rapidly coinciding with the consumption of surface oxygen and is integrated into a single band at 2060 cm^{-1}, the characteristic frequency of CO adsorbed on an oxygen-free Ru{0001} surface.[159]

The coadsorption of CO and O on a Ru{0001} surface was investigated under UHV using time-resolved IR reflection spectroscopy, thermal desorption mass spectroscopy, and LEED.[160] Time-resolved IR measurements were also used in the investigation of the formation and decomposition of formate species on clean and on potassium-modified Ru{0001} surfaces.[161] Characteristic differences were found in the spectra between the formate on the clean surface and that on the potassium-modified surface, showing the formation of potassium formate on the modified surface. Time-resolved spectra obtained during the thermal decomposition of formate revealed that the C–H and C–O bond cleavage reactions occurred simultaneously on the clean surfaces, resulting in the production of equal amounts of CO and CO_2. The presence of potassium suppresses the C–H bond cleavage, leaving CO and OH as the main decomposition products.

Ruett-Robey *et al.* combined a pulsed supersonic dosing apparatus with an FTIR spectrometer (see Fig. 10), capable of scan times of about 5 ms, to examine CO adsorbed on a periodically stepped Pt{111} surface. Using this technique, they monitored the migration of CO from terrace to step sites and obtained the hopping rate of CO on the {111} terraces.[162–165] The investigation was extended to a surface with higher step density covered with varying number of CO molecules, and the diffusion barrier (4.0 ± 0.7 kcal/mol) and prefactor ($10^{9.2\pm1.2}$ s^{-1}) for microscopic surface hopping of CO/Pt{111} were determined.[36] Typical time-resolved IR spectra recorded in this work are shown in Fig. 27. The migration of CO from terrace to step sites can be visualized from this figure.

The vibration of the bond formed between adsorbates and surface atoms has a frequency in the far infrared region. Since direct information on the adsorption strength is obtained from the low-lying vibration, the vibrational spectra in the far infrared region have been an object of investigation with high-resolution electron-energy loss spectroscopy.[166] Infrared reflection spectroscopy, on the other hand, has seldom been employed in the observation of the far infrared region, because the intensity of the thermal emitter in the source of IR spectrometers is weak in the long wavelength region and infrared emission has sometimes been observed in the investigation of this region.[167]

Synchrotron radiation, however, has higher brightness than the thermal emitter in the long wavelength region and can be used as a far infrared source.[29] Hirschmugl *et al.* measured reflection spectra from CO adsorbed on

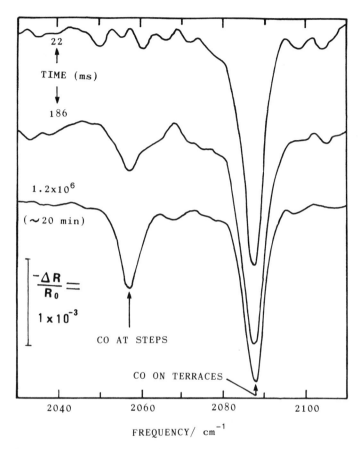

Figure 27. Time-resolved IR spectra showing the surface migration of CO from terrace to step sites. Time (in ms) is referenced to the arrival of CO at the surface. Temperature = 117 K, θ = 0.009 ML, RES = 4 cm^{-1} [Reprinted from J. E. Reatt-Robey, D. J. Doran, Y. J. Chabal, and S. B. Christman, *J. Chem. Phys.* **93**, 9113 (1990)].

Cu{100} in the region of 200–500 cm^{-1} using synchrotron radiation, and located features due to the Cu–CO stretching and the CO frustrated rotation modes.[30] The Fano-type line shape of the latter feature is attributed to the coupling to substrate electronic transitions. Hoffman *et al.* investigated the interaction of oxygen with potassium adsorbed on a Cu{100} surface from reflection spectra in the region of 50–600 cm^{-1} utilizing likewise synchrotron radiation. The formation of potassium super oxide, its decomposition at 600 K, and the simultaneous formation of Cu_2O were observed.[31] The promotion of oxidation of Cu by potassium was also found in this work.

In the observation of far infrared reflection spectra without a synchrotron radiation source, Hoge *et al.* used a Ge:Cu detector and observed a feature due to Pt–CO stretching vibration from a CO/Pt{111} system and its change as a function of increasing hydrogen exposure.[168] The detection of the same feature was also performed using a silicon bolometer detector.[169] Wang *et al.* observed reflection spectra of oxygen adsorbed on a Ag ribbon for the region 280–1700 cm^{-1} using the apparatus shown in Fig. 9. Three forms of oxygen species were identified in the measurements. Two of these are molecular forms and are standing up on the surface or lying parallel to the surface. The third form, atomic oxygen, has an Ag–O stretching vibration at 351 cm^{-1}.[170] Atomic oxygen adsorbs onto the silver surface via Langmuir kinetics, and the heat of adsorption differs by 12 kcal/mol from the desorption activation energy. The difference is attributed to the presence of an activation barrier to dissociative adsorption. The standing-up molecular oxygen is identified as a peroxo (O_2^{2-}) species and the lying-down molecular oxygen as the precursor to dissociative oxygen adsorption.[27,171] They also observed a band at 430 cm^{-1} due to the metal-carbon stretching vibration for CO on a silver foil.[172]

Adsorption isotherms of CO on a Cu{100} surface were determined using IR reflection spectroscopy at elevated pressures and temperatures.[173] In this work the absorption spectra of gas-phase CO were recorded separately. The spectrum of adsorbed CO was found by subtracting the gas-phase reference spectrum from the total spectrum. The isosteric heats of adsorption obtained in this work are in agreement with the previous results from the same system but at low pressures and temperatures.

Infrared reflection spectroscopy can also be used for investigating desorption and decomposition of CO adsorbed on a Ni{111} surface during electron bombardment.[174] The obtained spectra show that three-fourths of terminal CO is desorbed by electron bombardment before the beginning of the dissociation and desorption of bridged CO. The dissociation cross section of bridged CO is found to be larger than the desorption cross section, in contrast to the assumption hitherto accepted. The total cross section for bridged CO depletion is found to be one-tenth that of terminal CO.

Infrared reflection and absorption spectroscopy has also been used for studying the desorption of CO chemisorbed on an NiO film on Ni{111} and on a porous NiO/SiO_2 catalyst during UV photo-irradiation.[175] The obtained data for a photon energy of 3.82 eV show the presence of two kinds of adsorbed CO species having, respectively, photodesorption cross sections of ~2×10^{-18} and ~2×10^{-19} cm^2. Under the irradiation of 3.82 eV photons, CO desorbs also from porous NiO/SiO_2, but the desorption proceeds more slowly due to the diffusion initiation in the pores.

Adsorption of a variety of species other than those mentioned so far has been studied with IR reflection spectroscopy. The IR cross section of the umbrella mode of NH_3 is enhanced as a result of chemisorption on Ru{0001}. The enhanced cross section, however, reduces drastically upon formation of hydrogen bonding with NH_3 in the overlayer.[176] The adsorption of D_2O and H_2O on Pt{111} was studied with IR reflection spectroscopy combined with UV-photoemission. The chemisorbed water molecule is bound with its oxygen atom to the metal surface. With an increase in surface coverage, the water molecule forms multimers and then grows into water clusters at 120 K.[177]

The adsorption of N_2 on Ni{111} was studied using IR reflection spectroscopy under steady-state conditions at 89–115 K. The obtained spectra are shown in Fig. 28. At very low coverage, a feature is seen at 2218 cm^{-1}, which is assigned to the N≡N stretching mode of a singleton N_2 adspecies. At intermediate coverages, two other N_2 stretching bands are seen at 2208–2212 and 2203–2204 cm^{-1}. These two bands are assignable to N_2 species on the perimeter of and inside the ($\sqrt{3} \times \sqrt{3}$) R30° islands. At high coverages, a single peak is

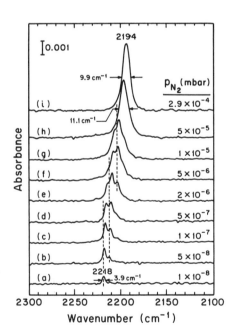

Figure 28. FTIR spectra of N_2 adsorbed on Ni{111} at 89 K as a function of N_2 pressure (100 scans, 2 cm^{-1} resolution) [Reprinted from J. Yoshinobu, R. Zenobi, J. Xu.and J. T. Yates, Jr., *J. Chem. Phys.* **95**, 9393 (1991)].

observed at 2194 cm^{-1} and is assigned to an in-phase collective vibration of an ordered ($\sqrt{3} \times \sqrt{3}$) R30° N$_2$ adlayer.[178]

The P–F symmetric stretching frequency of PF$_3$ chemisorbed on a Ni{111} surface shifts from 858 cm^{-1} to 916 cm^{-1} with an increase in coverage, showing strong lateral interaction between adsorbates, which is attributed to dipole-dipole coupling. At 273 K the band is sharp and symmetric in shape at both low and high coverages, indicating a uniform environment for PF$_3$ molecules. At intermediate coverages, the band is broad and asymmetric, indicating a wide distribution of intermolecular distances between chemisorbed PF$_3$. At low coverages the shape of the stretching band shows a strong temperature dependence, which is attributed to the change in population of higher-lying torsional states.[179]

2. Polarization and Wavelength Modulation Methods

2.1. Polarization Modulation Method

2.1.1. Background

Infrared spectra of species on metal surfaces are obtained by the oblique incidence method. The sensitivity of this method may be enhanced somewhat by a modification to be described below. However, when the observation is performed in the presence of absorbing gas-phase molecules, the signals from gas-phase species may give rise to a serious problem. The attenuation of the absorption of the solvent is essential for the observation of species on electrode surfaces. In observation at elevated temperatures, emission from the sample molecules has a substantial intensity and gives rise to the distortion of absorption bands in the low wavenumber region. At elevated temperatures, emission from the IR cell may cause degradation in the signal-to-noise (S/N) ratio of the obtained spectra.

Various techniques such as wavelength modulation, electrochemical modulation, polarization modulation, and its related technique of IR ellipsometric spectroscopy have been developed. This section deals with the polarization modulation method. The double-beam operation of the spectrometer has also been employed for the elimination of background absorption and was described in Section 1.2 of this chapter. The emission from the IR cell may be eliminated using a double-beam method, provided a high temperature cell is placed in the optical path of the reference beam.

When p-polarized IR light of unit strength is incident on a highly reflective metal surface in air, a standing electric field is generated on the surface and its component normal to the surface has an approximate strength $4\sin^2 \phi_1/n_2^3$ at the surface, as described in Section 1.1. At the same time, the standing field has a component parallel to the surface of strength about $4/k_3^2$.[4] On the other hand, the incidence of s-polarized light of unit strength gives rise to an electric field oscillating parallel to the surface having an approximate strength of $4\cos^2\phi_1/k_3^2$ at the surface.[4] Since the extinction coefficient, k_3, of highly reflective metals has a value in the range 15–130 in the infrared region, the strength of the oscillating field generated by the incidence of p-polarized light at 80° is a factor of 10^3 or more greater than that of s-polarized light incident at the same angle.

The intensity of IR absorption is proportional to the strength of the electric field that induces the absorption. It follows that when plane-polarized light is incident at a high angle on a highly reflective metal surface and the plane of polarization is rotated, the absorption intensity of the surface species changes in accordance with the rotation of the plane. Absorption of molecules in isotropic gas- and liquid-phase media, on the contrary, remain constant in intensity independent of the rotation of the plane of polarization. The phase-sensitive detection of the modulated signal therefore results in the elimination of the absorption due to gas- and liquid-phase molecules. Polarization modulation also eliminates non-modulated emission from the IR cell at elevated temperatures.

Bradshaw and Hoffmann constructed a polarization modulation apparatus using a rotating polarizer[180] and later using the alternate incidence of p- and s-polarized IR beams[181] for the elimination of water vapor absorption. They were able to detect an absorption band of CO adsorbed on Ru{0001} at a coverage of about 0.003 ML using this apparatus.[18] Blanke *et al.* combined a rotating polarizer with variable wavelength filters for the polarization modulation measurement and obtained a spectrum of a nine layer stearate film on a gold substrate in the 6-μm region, where water vapor exhibits very strong absorptions.[182] Spectra of CO and NO adsorbed on Pt foils were observed by the same researchers in the presence of gas-phase molecules at relatively high pressures.[183,184]

The oscillating electric field generated on a metal surface upon incidence of p-polarized light reduces in intensity with distance from the surface, but in contrast, the electric field arising from s-polarized light increases in intensity with distance from the surface. Since, in the polarization modulation method, the difference between the signal measured by p-polarized light and that by s-polarized light is observed, all the absorbing species present in a thin layer on the metal surface, whose thickness changes depending upon the wavelength of the incident light, are detected with this method. Consequently, when the polarization modulation method is utilized for the measurement of metal surfaces in vacuum (or in air), the species chemisorbed on a clean metal surface or those within and adsorbed on a thin film on the metal are observed, depending upon the condition of the sample. However, when this method is applied to the measurement of electrode surfaces, one often has difficulty distinguishing species adsorbed on the metal electrode from solute species present in the vicinity of the electrode.

The spontaneous emission from surface species at elevated temperatures can be eliminated by the use of the polarization modulation method. Figure 29 shows an IR feature of a thin aluminum oxide film on a gold substrate obtained with the polarization modulation method.[185] In this measurement, a rotating polarizer was placed in front of (I) or behind (R) the sample (i.e., the incident or the reflected light was modulated). When the incident light is modulated, the feature reduces only slightly in intensity upon increasing the temperature from 298 to 473 K, as can be seen in Fig. 29I, but in contrast, a substantial decrease in intensity is recorded in the measurement using the arrangement R (see Fig. 29R). Whereas the intensity of induced emission from the sample molecules is modulated in arrangement I, spontaneous emission is not modulated and is eliminated from the recorded spectrum. The above-mentioned slight decrease

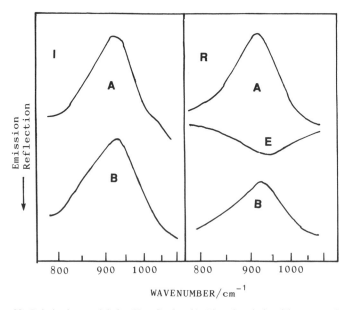

Figure 29. Polarization modulation IR reflection (A, B) and emission (E) spectra of a thin aluminum oxide film on a gold substrate. Infrared beam was modulated before (I) or after (R) the reflection at the sample surface. Sample temperature: (A) 298 K, (B) and (E) 473 K (adapted from Ref. 185).

in intensity with rise in temperature can be attributed to the increase in induced emission of the oxide film due to the rise in temperature.

Under the conditions of this measurement, the population in the second and higher excited states is very small and can be neglected. Let us then consider the transitions between the ground and the first excited states. The intensity of (induced) absorption and that of induced emission in the presence of radiation of frequency v, having a density $\rho(v)$ are, respectively, proportional to $N_0 B_{10} \rho(v)$ and $N_1 B_{01} \rho(v)$, where N_0 and N_1 are the number of molecules per unit volume in the ground and the first excited states, and B_{10} and B_{01} are absorption and induced emission coefficients, respectively. The intensity of induced emission relative to that of absorption is, therefore, N_1/N_0, because B_{01} equals B_{10}. With the Boltzmann distribution and assuming nondegenerate quantum states, one can estimate the increase in induced emission intensity upon increasing the temperature from 298 to 473 K. Calculation shows the increase in emission intensity is about 5% of the absorption intensity at 293 K, and is in agreement with the observed results shown in Fig. 29I.

Besides the induced emission, spontaneous emission is modulated in arrangement R, because the emission is polarized in the plane normal to the metal surface, as will be discussed in Chapter 4. In the obtained spectra, therefore, both the induced and the spontaneous emissions superimpose on the absorption band. The substantial intensity decrease shown in Fig. 29R can be attributed, for the most part, to the increase in spontaneous emission, because the increase in induced emission is insignificant.

Spontaneous emission spectra of aluminum oxide films can be observed using arrangement R with the light source removed. The obtained emission feature is shown in Fig. 29R. Addition of the emission feature E to the absorption band B gives a feature (not shown in this figure) that is in agreement with the feature A in the same figure. It is thus confirmed that the substantial decrease in intensity observed in the measurement using arrangement R arises from the increase in spontaneous emission. In this discussion, the feature E, spontaneous emission at 473 K, is used instead of the intensity increment of spontaneous emission, because spontaneous emission at 293 K is nearly equal in intensity to the increment of induced emission with rise in temperature from 298 to 473 K. The use of the polarization modulation method in arrangement I thus attenuates the observed intensity of emission to less than one-tenth that of the absorption in the wavenumber region higher than 800 cm^{-1} at temperatures lower than 500 K, and relatively undistorted reflection spectra are obtained with this method. The measurement of emission spectra may provide better results than the reflection measurement in the study of lower frequency regions at higher temperatures.

The polarization modulation method can also be utilized in the observation of thin films on semiconductor substrates such as Si and Ge. The reflectance of silicon changes by ΔR as a result of the introduction of a thin absorbing film on its surface, as shown in Fig. 7. Since, in contrast to the metal surfaces, the reflectance of the semiconductors changes appreciably in the presence of an absorbing thin film for s-polarized incident light, the use of the incident angle yielding the maximum value of $|\Delta R_p - \Delta R_s|$ instead of $|\Delta R_p|$ is important for obtaining good spectra.

Figure 7 shows that the difference $|\Delta R_p - \Delta R_s|$ increases at incident angles around 60° and 85°. However, the reflectance for p-polarized light is very low if an incident angle near 60° is used. In addition, the area of the film sampled by the incident IR beam increases steeply at high incident angles. The incidence of an IR beam at about 85° is, therefore, preferable in the observation of polarization modulation spectra of thin films on semiconductor substrates. Although the sensitivity of the polarization modulation method is low in the measurement of thin films on semiconductor surfaces in comparison with the measurement of metal surfaces, good spectra can be obtained for polymer films having thicknesses of the order of 10 nm. Note that the reflectance increases at the wavenumbers of the absorption bands of the film in this modulation measurement, and the spectrum obtained has the semblance of an "emission spectrum." The absorptions of gas-phase molecules and spontaneous emission from the sample molecules are also eliminated from the recorded spectra in the measurement of semiconductor surfaces.

2.1.2. Instrumentation

Polarization modulation may be carried out by placing in the optical path of the IR beam a mechanically rotating polarizer[33,180,182,186] or the combination of a photoelastic modulator and a fixed polarizer[34,187,188] before or after the sample. An alternative is the alternate incidence of p- and s-polarized IR beams, which has been used in combination with a grating monochromator.[176,189,190]

A variable wavelength filter, a grating monochromator, or an interferometer may be used for obtaining the modulation spectra. Variable wavelength filters are workable over the wavelength region between 2.5 and 14.5 μm and can be used in combination with any of the modulation systems mentioned above. The resolution of the filters is low, especially in the short wavelength region: the experimental halfwidth is ~30 cm^{-1} at 3.3 μm and ~7 cm^{-1} at 14.0 μm when used in combination with the narrow slit attachment. The filters provide high optical throughput, however, and very simple optical systems can be constructed with them. An example is shown in Fig. 30. This apparatus has a double modulation scheme. Infrared radiation from the light source is

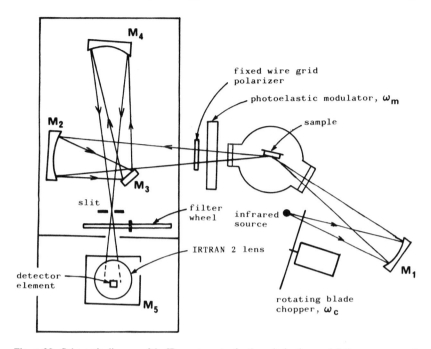

Figure 30. Schematic diagram of the IR spectrometer for the polarization modulation measurement [Reprinted from W. G. Golden, D. S. Dunn, and J. Overend, *J. Catal.* **71**, 395 (1981)].

chopped by a mechanical chopper (frequency ω_c). The radiation reflected from the sample is again modulated by a photoelastic modulator combined with a fixed polarizer (modulation frequency $2\omega_m$). The detector output of a signal modulated with a frequency ω_c and that doubly modulated with $2\omega_m$ and ω_c are separately demodulated through lock-in amplifiers tuned to reference frequencies ω_c and $2\omega_m$, respectively. The signal modulated with a frequency ω_c gives an output voltage proportional to $I_p + I_s$, where I_p and I_s are the intensity of the p- and s-polarized components of the radiation. After demodulation at $2\omega_m$ the doubly modulated signal gives an output, oscillating at a frequency ω_c, which yields, after additional demodulation at ω_c, a dc voltage proportional to $I_p - I_s$. The spontaneous emission signal from the sample is eliminated by the demodulation at ω_c. The ratio of these two outputs $(I_p - I_s)/(I_p + I_s)$ is recorded to give the spectrum of the surface species. The use of a photoelastic modulator capable of generating a much higher modulation frequency (>50 kHz) than that of the mechanical chopper is important for improving the S/N ratio.

In the reflection measurement, the IR beam is incident on the metal surface at or near the optimum angle. The background reflectance of highly reflective metals reduces to 0.7–0.8 at 2000 cm^{-1} when p-polarized light is incident at the optimum angle.[37] On the other hand, the background reflectance for s-polarized light has a value approaching unity at high angles of incidence. The reflectivity of reflecting mirrors also changes coincident with the rotation of the plane of polarization. The resulting imbalance of I_p and I_s gives rise to the appearance of absorptions of gas-phase molecules. In the double modulation measurement, the large imbalance may consume the better part of the dynamic range of the experiment, and the sensitivity may be reduced appreciably. This imbalance may be compensated for by inserting a thin parallel-sided plate of IR transparent crystal, such as KBr, into the IR beam at an appropriate angle. The transmittance through the plate for p- and s-polarized light depends upon the incident angle and wavelength, and one may reduce the difference between I_p and I_s to zero by changing the angle of the inserted plate with respect to the optical path.

An example of the combination of a grating monochromator with the alternate incidence of p- and s-polarized beams is shown in Fig. 31.[191] Infrared radiation from an IR source is divided into two beams, which are focused onto the fixed polarizers to produce p- and s-polarized radiation, respectively. The rotating mirror-blade sector recombines the radiation, and the pulses of p- and s-polarized radiation alternately hit the sample surface at a frequency of 140 Hz. The monochromator is inclined at 45° with respect to the planes of polarization of the IR beams so as to avoid the polarization-dependent variation in the reflectivity of the grating. A stationary polarizer, positioned with its passing axis at 45° with respect to the surface normal, may be used instead of the oblique mounting of the monochromator. An attenuator inserted into the optical path of the s-polarized beam is used to minimize the difference between I_p and I_s, i.e., the attenuator plays the role of the KBr plate mentioned above. The performance of the apparatus is illustrated by an example of a single monolayer of stearic acid on a silver substrate. The spectrum was obtained with a good S/N ratio, as shown in Fig. 32.

Interferometers are used in combination with a photoelastic modulator-fixed-polarizer assembly for polarization modulation measurements, because modulation with a frequency much higher than the interferometer modulation frequency is required.[24,192–195] Figure 33 shows an example of this optical arrangement. Infrared light from the interferometer is focused on a photoelastic modulator and then onto the sample substrate at grazing incidence. The reflected light is finally focused onto a detector. The doubly modulated signal from the detector is processed, in principle, in the same manner as described

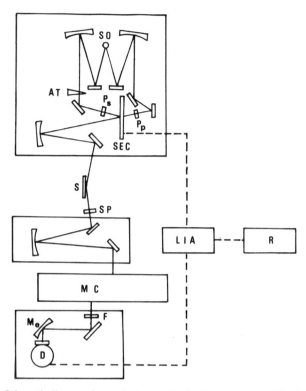

Figure 31. Schematic diagram of a polarization modulation IR spectrometer. SO = light source, AT = attenuator, P_s, P_p = polarizer, SEC = sector mirror, S = sample, MC = monochromator, F = filter, M_e = off-axis ellipsoidal mirror, D = InSb/HgCdTe dual detector, LIA = lock-in-amplifier, R = recorder, SP = stationary polarizer (see text) [Reprinted from A. Hatta, T. Wadayama, and W. Suëtaka, *Anal. Sci.* **1**, 403 (1985)].

above. Figure 34 shows the reflection spectra of CO adsorbed on a platinum foil produced using the configuration shown in Fig. 33.

One might expect that information from a wide wavelength region would be obtained at high resolution and high S/N ratio through the combination of an interferometer and a photoelastic modulator–polarizer assembly. However, this is not the case. The imbalance of I_p and I_s is frequency-dependent, and compensation over the entire spectral region cannot be attained, resulting in the imperfect elimination of the background absorptions. In addition, since the phase retardation induced by the photoelastic modulator is frequency-dependent, the plane-polarized light is obtained solely at a specific wavelength,

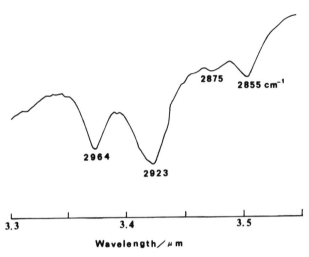

Figure 32. Infrared reflection spectrum of a single monolayer of stearic acid on a silver substrate in the C–H stretching vibration region [Reprinted from A. Hatta, T. Wadayama, and W. Suëtaka, *Anal. Sci.* **1**, 403 (1985)].

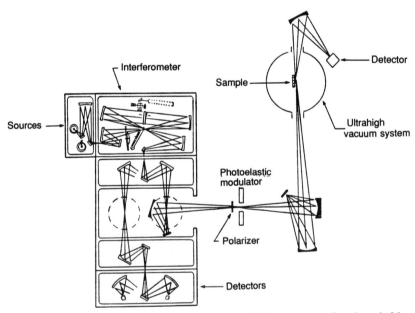

Figure 33. Schematic diagram of a polarization modulation FTIR spectrometer. One channel of the IBM Instruments IR/98 is used for ambient monolayer samples, and the other for the polarization modulation measurement of samples in ultrahigh vacuum [Reprinted with permission from W. G. Golden, D. D. Saperstein, M. W. Severson, and J. Overend, *J. Phys. Chem.* **88**, 574 (1984), © 1984 American Chemical Society].

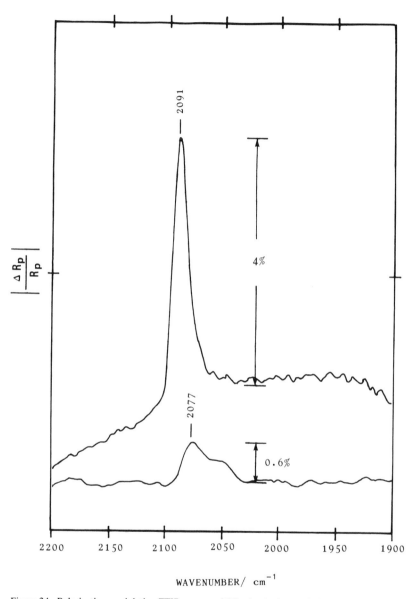

Figure 34. Polarization modulation FTIR spectra of CO adsorbed on a platinum foil at 300 K. The upper trace is the spectrum of saturation coverage of CO on the Pt foil and the lower one is the spectrum of low coverage of CO [Reprinted with permission from W. G. Golden, D. D. Saperstein, M. W. Severson, and J. Overend, *J. Phys. Chem.* **88**, 574 (1984), © 1984 American Chemical Society].

variable, however, with a change in the power applied to the piezoelectric driver. The photoelastic modulator is operated at a fixed power in each FTIR measurement, and the spectrum at a high S/N ratio is obtained in a limited spectral region in a single measurement.

2.1.3. Applications

2.1.3.1. Solid Surfaces in Absorbing Gas-Phase Molecules. Reaction intermediates and physisorbed species on a catalyst surface play important roles in the catalytic reaction on the surface. The observation of the behavior of these species should provide important clues for the better understanding of the reaction mechanism. Evacuation of gas-phase molecules, which is necessary for measurements with techniques using electrons or ions as probes, however, results in the removal of the weakly adsorbed and metastable species from the surfaces at relatively elevated temperatures where the reaction proceeds. Consequently, techniques that enable us to observe *in situ* solid surfaces under gas pressure at relatively elevated temperatures are required. Since the polarization modulation method eliminates the signals of absorption of gas-phase molecules and spontaneous emission from surface species, it is eligible, among others, for the *in situ* observation of physisorbed species and metastable reaction intermediates, as well as chemisorbed species.

Overend and his collaborators observed reflection spectra of CO adsorbed on Pt in the presence of gas-phase CO at a pressure of ~1.3×10^4 Pa and found a marked feature at around 2070 cm^{-1} without the overlapping of absorptions from gas-phase CO[183]; they also observed a feature at 1875 cm^{-1} of CO adsorbed at bridge sites in 6.7×10^3 Pa CO.[196] They also investigated the behavior of NO adsorbed on Pt by changing the pressure of gas-phase molecules and the substrate temperature.[184,197] The spectrum of adsorbed NO changes markedly depending upon the partial pressure of NO and the temperature, showing that NO adsorbed on Pt is in equilibrium with the gas phase at elevated temperatures.[197]

The oxidative decomposition of formic acid and formaldehyde on oxygen-covered copper surfaces was observed *in situ* in the presence of gas-phase molecules with the polarization modulation method.[186,198] The reflection spectra from the copper surface in formaldehyde vapor at a pressure 2.6×10^2 Pa show the presence of physisorbed formaldehyde, uni- and bidentate formate ions, and a species with spectral features at 980 and 2880 cm^{-1} at room temperature. Physisorbed formaldehyde is oriented with its C=O bond parallel and the CH$_2$ plane slightly inclined to the metal surface and is desorbed upon evacuation at room temperature. The bands at 980 and 2880 cm^{-1} disappeared as a result of substrate heating to 350 K in vacuum, while the features due to bidentate formate ion increased in intensity. These two bands may be assigned to a reaction intermediate, dioxymethylene,[199] adsorbed with its C_2 axis normal to the metal surface.[198]

The hydrogenation of methyl acetoacetate over Raney nickel modified with an optically active hydroxy or amino acid yields either of the optical isomers of methyl β-hydroxybutyrate depending upon the optical activity of the acid used as the modifying reagent.[200] It is assumed that the orientation of the modifying reagent adsorbed on the catalyst and the interaction of substrate molecule with the modifying reagent play important roles in determining the configuration of the reduction product.

Reflection spectra of ethyl acetoacetate adsorbed on nickel surfaces with or without modifying reagent D-α-alanine were observed in the presence of gas-phase acetoacetate molecules at a pressure of 1.3×10^2 Pa.[201] The spectra obtained are shown in Fig. 35. Spectrum A comes from ethyl acetoacetate molecules adsorbed on an unmodified nickel substrate. The tracing B is a difference spectrum which is obtained by subtracting the spectrum of the modifying reagent molecules adsorbed on a nickel substrate from the spectrum taken after admission of acetoacetate vapor.

Spectrum A shows two features at 1238 and 1747 cm⁻¹, which are as-

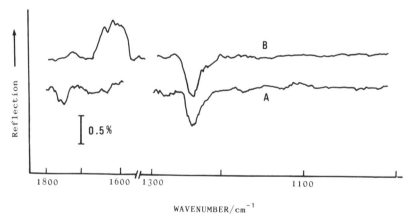

Figure 35. Infrared reflection spectra of ethyl acetoacetate adsorbed on polycrystalline nickel (partial pressure of acetoacetate = 133 Pa). (A) Spectrum of acetoacetate adsorbed on a nickel substrate without modifier, and (B) spectrum of acetoacetate on a nickel surface modified with D-α-alanine (adapted from Ref. 201).

signed to C–O–C antisymmetric stretching and C=O stretching vibrations of the ester group in keto form acetoacetate, respectively. The feature at 1747 cm^{-1} disappears in Spectrum B, showing the change in orientation of adsorbed acetoacetate molecules caused by the presence of D-α-alanine. The positive-going feature centered at 1618 cm^{-1} indicates the intensity decrease of the NH$_3^+$ degenerate deformation and/or the CO$_2^-$ antisymmetric stretching bands of alanine as a result of coadsorption of acetoacetate molecules. The evacuation of acetoacetate vapor gives rise to the recovery in intensity of the 1618 cm^{-1} band and the disappearance of the 1238 cm^{-1} band. The intensity decrease of the former band upon adsorption of acetoacetate implies, therefore, that the geometry of adsorbed D-α-alanine changes as a consequence of the interaction with acetoacetate.

Figure 36 shows reflection spectra of a MoO$_3$ thin film catalyst on a gold substrate taken at 473 K before and after exposure to methanol vapor at a pressure of 5.3×10^2 Pa.[202] The exposure to methanol causes a steep decrease in the intensity of a feature at 990 cm^{-1}, which is assigned to the Mo=O stretching vibration.[203] On the other hand, a broad feature centered at 870 cm^{-1} due to the Mo–O–Mo vibration[203] decreases in intensity slowly. This indicates that methanol reduces Mo=O groups preferentially. The appearance of features at 940, 1080, and 2920 cm^{-1} (not shown in this figure) in the difference spectrum E implies the presence of metastable species on the catalyst. The former feature at 940 cm^{-1}, that also appears in the spectrum of MoO$_3$ exposed to hydrogen,[198] may be assigned to the Mo(V) = O vibration, and the latter two features may be assigned to the C–O and C–H stretching vibrations of surface methoxy group, respectively.

Polarization modulation IR spectroscopy has been applied also to the *in situ* observation of a Pt/TiO$_2$ (anatase) photocatalyst in a methanol–water mixture vapor under irradiation of UV light.[204] The spectrum obtained shows the features assignable to methoxy groups adsorbed on anatase. The features persist under UV irradiation as long as methanol vapor is present in the reaction cell, but disappear upon the introduction of water vapor subsequent to the evacuation of methanol. This result supports the deduction that the hydrogen evolution from a methanol–water–Pt/TiO$_2$ system proceeds through the decomposition of the methoxy group adsorbed on anatase by water.[205]

As mentioned in the beginning of this section, polarization modulation IR spectroscopy provides information about all of the absorbing species in the proximity of the metal surface, in contrast to surface-enhanced Raman scattering, which results from short-range mechanisms, and enhanced infrared spectroscopy, which owes to the enhanced local electric field generated at rough silver and some other metal surfaces.

Wavelength/μm

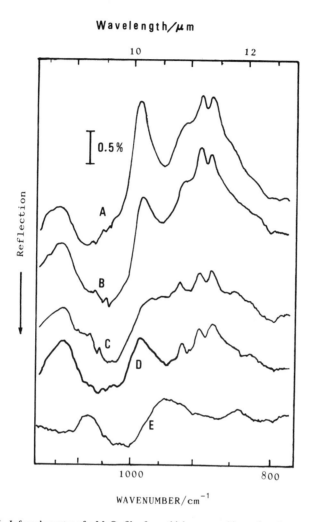

Figure 36. Infrared spectra of a MoO₃ film 3 nm thick measured in methanol vapor at 473 K. (A) Before introduction of methanol. (B) spectrum recorded immediately after the admission of methanol, (C) spectrum taken 2 h after the admission, (D) after evacuation of C, and (E) difference spectrum between spectra D and C [Reprinted from T. Wadayama, T. Saito, and W. Suëtaka, *Appl. Surf. Sci.* **20**, 199 (1984)].

Knoll *et al.* investigated monolayer assemblies of cadmium arachidate molecules deposited by the Langmuir–Blodgett technique on silver surfaces with polarization modulation IR and surface enhanced Raman spectroscopy.[206] Evaporation of island Ag films on the top of the arachidate layers brings about marked changes in Raman spectra, but in contrast, the infrared spectra are almost unaffected by the evaporation. These results have been explained with an assumption that surface-enhanced Raman spectra yield information about disordered molecular species present in narrow regions where electromagnetic fields are enhanced by a plasmon resonance or some other mechanism, while IR spectroscopy probes all molecules in the assemblies. Since the number of disordered molecular species is much smaller than the ordered species, the overwhelming infrared signals from the latter conceal those of the former from the obtained spectra.

Vapor deposition is widely used for the fabrication of thin semiconductor and dielectric films, which are important as microelectronic devices. *In situ* observation of the films growing during the deposition process should provide valuable information for optimizing the process of deposition to yield high quality layers. Polarization modulation IR reflection spectroscopy is quite feasible for the *in situ* observation of thin films growing in reactant gas. Koller *et al.* observed *in situ* the growth of SiO_2 films on HgCdTe, silicon, and aluminum substrates during plasma-enhanced chemical vapor deposition using a combination of an interferometer and a photoelastic modulator.[207]

The growth of hydrogenated amorphous silicon (a–Si:H) on a gold substrate was observed *in situ* in disilane gas in the course of the photochemical vapor deposition process. The spectra obtained show that a highly hydrogenated species of $-SiH_3$ is predominant in the film at the initial stage of deposition at 293 K, but its concentration begins to decrease when the film thickness increases beyond 1.5 nm.[208] The features due to the highly hydrogenated species decrease in intensity accompanying simultaneous evolution of hydrogen gas upon exposure to tungsten hexafluoride gas, indicating that the hydrogenated species is present for the most part in the outermost layer of the film.[209a]

Infrared spectra of the a–Si:H films show the presence of Si–O bonds when measured in air after deposition, and the agent giving rise to the bond has been the subject of controversy. Figure 37 shows the change in spectrum of a vapor deposited a–Si:H film upon exposure to air. The feature at 1175 cm^{-1} assignable to the Si–O vibration develops after the introduction of air into the reaction chamber, which indicates that the Si–O bond is not formed in a reaction of the growing film with residual gas in the chamber.[208]

In situ IR observation of SiN_x films growing through photochemical vapor

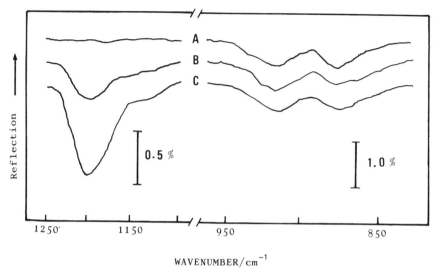

Figure 37. Oxidation of a hydrogenated amorphous silicon film deposited at 293 K. (A) Spectrum recorded before introduction of air, (B) spectrum taken 30 min after the introduction of air, and (C) 2 h after the introduction (adapted from Ref. 208).

deposition has revealed that reactive gas-phase species produce, upon arrival at the solid surface, a metastable layer that converts into the stable silicon nitride network under UV irradiation.[45] Another *in situ* observation indicated that the conversion of the metastable layer into the stable network took place also by the dry etching of the former with F_2 gas, and the conversion was accompanied with evolution of SiF_4 and H_2.[209b]

Spontaneous chemical deposition of a–Si:H(F) films through a gas-phase reaction of SiH_4 with F_2 was investigated at a temperature range 373–523 K with *in situ* observation of polarization modulation IR spectra.[210] The spectra obtained show the presence of fluorinated silicon species besides hydrogenated species. Fluorinated silicon groups such as $-SiF_3$ and $-SiF_2-$ are predominant in the film at the beginning of deposition. The rise in deposition temperature results in a remarkable decrease in hydrogenated and fluorinated silicon species, showing the acceleration of silicon network formation due to the increase in temperature.

The polarization modulation method has been applied to the observation of thin films on semiconductor substrates. Figure 38 shows a reflection spectrum of a poly(vinyl acetate) film 10 nm thick on a silicon wafer obtained with

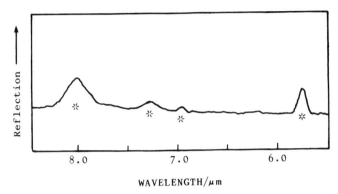

WAVELENGTH/μm

Figure 38. Infrared reflection spectrum of a poly(vinyl acetate) film of 10 nm in thickness on a silicon wafer measured at an incident angle of 85° [Reprinted from A. Hatta, T. Wadayama, and W. Suëtaka, *Anal. Sci.* **1**, 403 (1985)].

the polarization modulation method.[191] All of the principal absorption bands (denoted with an asterisk) in the region of this figure are clearly seen in the spectrum. The sensitivity of the polarization modulation method is low in comparison with the multiple internal reflection method as far as thin films on semiconductor substrates are concerned. However, it should be useful for the observation of thin films on semiconductor surfaces at elevated temperatures, where the semiconductor is opaque for the free carrier absorption.

2.1.3.2. Electrode Surfaces. In the investigation of electrode processes, electrochemical measurements have been widely used for many years. However, these measurements provide no direct information on the identity of species at the electrode/electrolyte interface, and optical techniques using visible and near ultraviolet light have been introduced for the investigation of the interface.[211,212]

The optical techniques still proved unsatisfactory for obtaining information on the structure of species at or near the electrode surface as well as information on the orientation of species adsorbed on the electrode. Raman spectroscopy, then, was used to identify surface species, and electrochemical and polarization modulation infrared spectroscopy was employed to obtain this kind of information.[213]

The polarization modulation method is somewhat lower in sensitivity than electrochemical modulation in the observation of electrochemical processes on electrodes. However, the shape, peak frequency, and intensity of the features cannot be determined by electrochemical modulation spectroscopy, because the difference between the overlapping features of comparable intensity is measured by the electrochemical modulation method. Furthermore, electrochemical modulation approaches are inapplicable to the investigation of irreversible processes on electrodes. Polarization modulation spectroscopy, therefore, is used as a complementary technique to electrochemical modulation spectroscopy for the vibrational investigation of electrode surfaces.

In the external reflection IR measurement of electrode surfaces, one faces the problem of intense absorption by the solvent. To obtain adequate transmission of IR light, the thickness of the solution layer must be a few micrometers or less. For this reason, the optical arrangement shown in Fig. 39 was devised. A trapezoidal prism window of silicon, ZnSe, CaF_2, or KRS 5 is used for collimated IR light. The incident light hits the prism window normal to the plane so that reflection at the prism-air interface is the same for p- and s-polarized light. When the incident light is convergent, a hemicylindrical window is used. The working electrode is attached to the end of a glass tube extension of a syringe plunger or to a copper rod coated with teflon. The

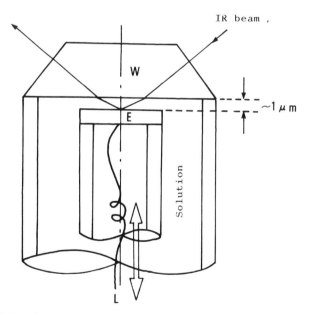

Figure 39. Schematic representation of a thin solution layer cell for the infrared measurement of electrode surface. W = IR transparent window, E = working electrode, and L = conducting wire.

above-mentioned thin layer of solution is formed by pushing the syringe plunger so that the working electrode comes into contact with the window. An example of the entire electrochemical cell is shown in Fig. 40.[214]

Russell *et al.* were the first to apply the polarization modulation method to make *in situ* observations of an adsorbed species on an electrode surface.[215] Figure 41 shows reflection spectra of CO adsorbed on Pt in 1-M $HClO_4$ saturated with CO gas. The peak frequency of the feature shifts to higher frequencies with a positive change in the electrode potential. The tracing B of this figure is the difference between the spectra iii-A and i-A, and is in perfect agreement with spectrum C, which is the electrochemical modulation spectrum of CO adsorbed on Pt in 1-M $HClO_4$ saturated with CO gas modulated between 0.05 and 0.45 V. The agreement manifests the significance of the polarization modulation method in the investigation of electrode surfaces, since the spectra at the two potential limits cannot be determined with the electrochemical modulation method. For instance, the two features shown in Fig. 41D could give an electrochemical modulation spectrum similar to spectrum C.

Although Fig. 41 shows only the feature due to terminally bound CO on a Pt electrode, CO also adsorbs on bridge sites on the platinum electrode, as is the case with clean Pt surfaces in ultrahigh vacuum.[216] The change in electrode potential gives rise to a shift in the feature due to bridge-bonded CO as well.

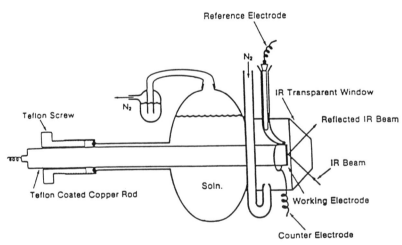

Figure 40. Schematic drawing of an electrochemical cell used for the external reflection IR measurement [Reprinted from K. Chandrasekaran and J. O'M. Bockris, *Surf. Sci.* **175**, 623 (1986)].

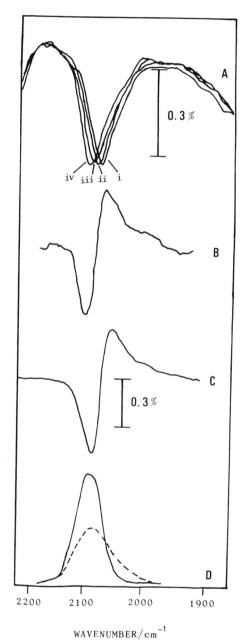

Figure 41. (A) Polarization modulation IR reflection spectra of CO on Pt in 1 mol/l HClO$_4$. The electrode potential was held constant at (i) 0.05, (ii) 0.25, (iii) 0.45, and (iv) 0.65 V (NHE). (B) Difference spectrum between a A (iii) and A (i). (C) electrochemical modulation IR spectrum of CO on a Pt electrode in HClO$_4$ (1 mol/l) modulated between 0.05 and 0.45 V. (D) hypothetical spectra which could produce a difference spectrum similar to that shown in (C) [Reprinted with permission from J. W. Russell, J. Overend, K. Scanlon, M. Severson, and A. Bewick, *J. Phys. Chem.* **86**, 3066 (1982). © 1982 American Chemical Society].

The change in the electrode potential results, at the same time, in the interchange of adsorption sites.[217]

The shift in the peak frequency with change in the electrode potential, which is seen in Fig. 41A, can be explained by the back-donation of metal electrons into the antibonding $2\pi^*$ orbitals of the adsorbed CO molecules or the vibrational Stark effect.[218,219] The lateral interaction changing with surface coverage, mentioned in Section 1.4.2, may be ruled out in this case as a dominant mechanism, because no change was observed in band shape and intensity. In the former mechanism, electrons are back-donated from partially filled d-orbitals of metal to $2\pi^*$ orbitals of the adsorbate. As the electron density of surface metal atoms decreases with the positive shift of the electrode potential, the back-donation into the antibonding $2\pi^*$ orbitals of the adsorbed CO molecules reduces correspondingly, resulting in an increase of the stretching force constant of the C–O bond.[220,221] The change in the electric field in the double layer may bring about the shift in energy level of the adsorbate antibonding state, thus affecting its occupancy by the metal electrons. Holloway and Nørskov[222] calculated the occupancy of this orbital of CO adsorbed on Ni as a function of the potential and deduced a linear shift of 30 cm^{-1}/V in the C–O stretching frequency.

The Stark effect mechanism involves an interaction of the electric field across the double layer with the highly polarizable electrons of the adsorbed CO molecules. Adsorbed CO molecules have their axial orientation parallel to the applied electric field. As a consequence, adsorbed molecules with a dynamic dipole moment have a first order Stark effect. The change in vibrational frequency by an electric field, E, acting on the molecule can be estimated from the Stark tuning rate $d\omega/dE$ (ω = vibrational frequency). Lambert calculated a Stark tuning rate of -8.6×10^{-7} cm^{-1}/(V/cm) for ^{12}C^{16}O.[223] Since an electric field of the order of 10^7 V/cm may be expected to exist in the layer of adsorbed species, a significant shift of the C–O stretching frequency of adsorbed CO will result. The stretching vibrational frequency of cyanide ion adsorbed on Ag, Au, or Cu electrodes also shows a shift with the change in electrode potential that is similar to that of CO adsorbed on Pt electrodes. Molecular orbital calculations suggest that the chemical bonding of CN$^-$ to the Cu surface is ionic without back-donation from the metal d-orbitals to the π-levels of CN$^-$. The origin of the potential dependence of the C–N stretching frequency is also ascribed to a Stark effect, though the detailed mechanism is different from the CO-metal electrode system.[224]

The polarization modulation method has some drawbacks in the observation of electrode surfaces. Since absorbing species present within a certain distance of the surface of the metal electrode are measured by this method, as described in Section 2.1.1 of this chapter, signals from the solute in bulk solution are recorded in the spectra together with those from species adsorbed on the electrode surface. Discrimination between signals of adsorbed species and those in solution is difficult, particularly when ions or molecules of complex structure are present on the electrode. This is also the case with the electrochemical modulation method, where the modulation of electrode potential induces the change in concentration of solute species in proximity to the electrode. The observation of reflection spectra using s-polarized IR light is often used to distinguish the adsorbed species from those in solution.[225]

As the IR light traverses the solution layer in the polarization and electrochemical modulation methods, its intensity is attenuated substantially by immense absorptions of the solvent, and certain features of the surface species are obscured by absorptions of the solvent molecules present near the surface. Furthermore, thin layer cells used in these modulation techniques may give rise to a serious problem, because the electric current is concentrated on the edge of the electrode. The interference effect may cause an additional serious problem. Signals of adsorbed species may change in intensity, and the band shape of the signals may be distorted severely as a result of the interference effect.[226]

Surface-enhanced Raman spectroscopy derived from the chemical mechanism (Chapter 6) and enhanced infrared absorption spectroscopy in the Kretschmann configuration (Chapter 3) appear to be superior substitutes for these modulation methods, because they have an advantage that makes the use of thin layer cells unnecessary: Raman signals of the solvent are much weaker than the infrared absorptions, and the observation of enhanced infrared spectra can be made with an internal reflection arrangement. Furthermore, only the signals of species directly bound to the electrode surface are enhanced in chemically enhanced Raman scattering. In the Kretschmann configuration, the enhanced surface electromagnetic field that yields enhanced infrared absorptions is confined to the closest vicinity of the metal surface. As a result, the signals of solute species in the bulk solution remain very weak in either of the techniques. However, the use of an electrode of particular metals having a rough surface is necessary for obtaining the enhanced Raman and IR spectra. Furthermore, only particular adsorbates give rise to chemically enhanced surface Raman spectra, and only IR signals from species within hollows in the rough surface of the electrode are enhanced by the localized EM field. For these reasons, the polarization modulation method is important for the *in situ* observation of electrode surfaces, notwithstanding the above-mentioned drawbacks.

The oxidation of CO adsorbed on a smooth platinum electrode in acidic solution was investigated by means of polarization modulation IR reflection spectroscopy.[216,227] The intensity, peak frequency, and bandwidth of the IR band of CO terminally bound to the electrode, as well as those of evolved CO_2, were measured as a function of potential. Whereas the oxidation of CO adsorbed at 0.4 V (NHE) in the double-layer potential region proceeds mostly at the edges of CO islands, the oxidation of CO adlayer formed at 0.05 V in the hydrogen region occurs randomly throughout the layer.

Since cyanide ion plays an important role in the electroplating process of several metals such as Au, Ag, Cu and its alloys, Zn, and Cd, its behavior on metal electrodes is quite important. The polarization modulation method has been applied to the *in situ* observation of cyanide ions on metal electrodes, though the intensity of adsorbed cyanide ions is one order of magnitude smaller than that of CO. For example, adsorption of cyanide ion on a polycrystalline gold electrode was observed *in situ* with the polarization modulation method.[225] A feature of 2105 cm^{-1} is attributed to the C–N stretching vibration of the terminally bound cyanide ions. At –0.7 V (Ag/AgCl) and more positive potentials, a new feature develops at 2146 cm^{-1} which is assigned to the $Au(CN)_2^-$ complex ion in the solution phase. The C–N stretching vibration of adsorbed cyanide ion shows a small shift with change in coverage at a fixed potential of –1.0 V (Ag/AgCl). Measurements using isotopic mixtures of ^{12}CN and ^{13}CN show that the shift is attributable to the change in lateral interaction between adsorbed cyanide ions. Cyanide ion adsorbed on a polycrystalline copper electrode shows a C–N stretching band at around 2090 cm^{-1}, which shifts depending upon the electrode potential in dilute cyanide solutions. In concentrated solutions, copper cyano complex ions of $Cu(CN)_4^{3-}$, $Cu(CN)_3^{2-}$, and $Cu(CN)_2^-$, whose peak frequencies are potential-independent, have been identified as solution species. The relative concentrations of the solution species are controlled by chemical equilibria and the free Cu^+ ion concentration which is, in turn, potential-dependent.[228]

The polarization modulation method has also been applied to the investigation of electrode surfaces in nonaqueous solutions. Chandrasekaran and Bockris[229] investigated electrochemical reduction of CO_2 on a Pt electrode in acetonitrile with *in situ* infrared observation of the electrode surface. The results obtained show that the reaction involves adsorption of CO_2 and CO_2^- anion radical on the electrode, with the surface concentration of the latter species increasing with the cathodic shift in the electrode potential. The mechanism of the reduction reaction is discussed.

The use of a clean and well-defined single crystal surface is indispensable in the full analysis of processes that take place on electrode surfaces, as is the case in the investigation of adsorption on solid surfaces in vacuum. However, clean and well-defined single-crystal surfaces have been used as electrodes only recently because of the difficulty in eliminating contamination and in assessing clean surfaces. Clavilier *et al.* first applied the linear potential sweep voltammogram of hydrogen adsorption-desorption on a Pt{111} surface to the evaluation of the clean surface of Pt single crystals prepared by a novel anneal-quenching technique.[230] In an investigation with LEED and AES measurements of a single crystal Pt electrode subjected to the voltammetric experiment, Aberdam *et al.* confirmed that the above linear sweep voltammogram does correspond to the hydrogen adsorption-desorption process on a clean, well-defined Pt{111} surface.[231]

Furuya *et al.* observed *in situ* the oxidation of CO adsorbed on Pt{111} in 0.5-M sulphuric acid with polarization modulation IR reflection spectroscopy. Bridge-bonded CO was not detected in this work, irrespective of the potential of CO adsorption, in contrast to the measurements with polycrystalline Pt electrodes. When CO was adsorbed at 0.05 V (NHE), the terminally bound CO on Pt {111} showed behavior in the oxidation proces different from that of CO adsorbed at 0.6 V. The difference in behavior may be attributed to the difference in the number of vacant sites which play an important role in the oxidation of the terminally bound CO. Reconstruction of the Pt{111} surface induced by a potential excursion into the oxide formation region gives rise to a remarkable increase in intensity of the feature due to terminally bound CO, in addition to the shift of the band peak to a higher frequency, but this has little effect on the electro-oxidation of CO.[232] The behavior of CO adsorbed on stepped Pt surfaces was also studied with the IR reflection method.[233]

The mechanism of corrosion inhibition in wet systems by organic molecules has been investigated extensively with IR reflection spectroscopy. However, most of the investigations have been carried out through *ex situ* observation of protective films formed on the metal surfaces by the use of inhibitors. *In situ* observation of the behavior of inhibitor molecules on metal surfaces should provide information valuable for better understanding of the mechanism of corrosion inhibition.

Ito and Takahashi have employed polarization modulation IR reflection spectroscopy for the *in situ* observation of copper surfaces in methanol and in water containing HCl and benzotriazole (BTA) as an inhibitor.[234] The film formed on a copper substrate in methanol has a structure similar to $Cu_2Cl_6(BTA)_6$ crystal. The polymer films formed in aqueous solutions exhibit reflection spectra similar to the films formed in methanol, and the increase in thickness is proportional to the time of contact with the corrosive environment. Measurements of BTA adsorbed on metal electrodes at various potentials may provide information valuable for the practical use of the inhibitor.

Vibrational spectra of the interface between pure pyridine or its aqueous solution and an oxidized aluminum surface were measured with a tunable diode laser IR spectrometer.[34] A new peak appears near 1455 cm^{-1} (in addition to the features of liquid-phase species) and is assigned to pyridine adsorbed at Lewis acid sites on the oxidized aluminum surface.

Polarization modulation IR reflection spectroscopy has been applied to the *in situ* observation of the semiconductor-electrolyte interfaces. Chandrasekaran and Bockris have studied adsorption of ions and neutral molecules on *p*-silicon and *p*-gallium phosphide electrodes in aqueous and nonaqueous solutions using the modulation method in combination with subtractively normalized interfacial IR spectroscopy.[214] In this measurement, strong features of the solvent appearing in the observed region were eliminated by the subtraction of spectra recorded at potentials where no adsorption of solute species is expected. The details of these results will be shown in Section 3 of this chapter.

2.1.3.3. Clean Surfaces. The polarization modulation method was first utilized for the elimination of the strong water vapor absorptions in the observation of CO adsorbed on clean Pd surfaces.[180,181] When this method was used in combination with a baseline subtraction, a noise level of ~0.02% was attained at a resolution of 3.5 cm^{-1}. Reflection spectra of CO adsorbed on Ru{0001} were obtained with this procedure at a high sensitivity, as shown in Fig. 42, where the C–O stretching band from less than 0.3% of monolayer of adsorbed CO can easily be detected.[18]

The adsorption of CO and ^{12}CO/^{13}CO mixtures on low index faces of single crystals of transition metals such as Pd, Ru, and Pt has been studied extensively by the polarization modulation method.[18,108,235,236] The shift in absorption peak with coverage was measured in combination with LEED observations, and the origin of the shift as well as the nature of the adsorption site were discussed. The modulation method was also employed in a dynamic study of isosteric phase transformation of a CO adlayer on Ni{100},[155] as mentioned in Section 1.4.2 of this chapter.

The effect of preadsorbed atoms such as C, N, O, S, and Cl on the adsorption of CO on Ni{100} has been investigated with temperature programmed desorption, polarization modulation IR spectroscopy, LEED, and AES. Whereas CO is adsorbed at bridge and on-top sites on the surfaces with adlayers of C, N, and Cl at 170 K, only bridge-bonded CO is present on a surface with an O or S adlayer. Above 300 K the adsorption sites changed except for the carbon preadsorption. The change observed has been explained by assuming the reconstruction of the p(2 × 2)X adlayer into islands of c(2 × 2)X and clean surfaces. The diffusion of carbon atoms into the bulk from a p(2 × 2)C adlayer was also postulated.[237] Similar effects were seen on Ni{111}.[121,122]

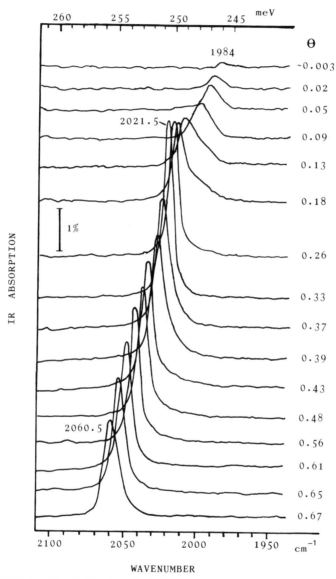

Figure 42. Infrared band arising from the C–O stretching vibration of CO adsorbed on Ru{0001} as a function of increasing coverage. Temperature of adsorption and measurement = 200 K [Reprinted from H. Pfnür, D. Menzel, F. M. Hoffmann, A. Ortega, and A. M. Bradshaw, *Surf. Sci.* **93**, 431 (1980)].

In addition to CO, the adsorption of water and other compounds on low index surfaces of ruthenium and nickel single crystals has been investigated with polarization modulation spectroscopy. Water molecules adsorbed on Ru{0001} exhibit at 85 K a very broad feature due to the O–H stretching vibration even at low coverages, implying that the hydrogen-bonded clusters are formed from the initial stage of adsorption. The cluster formation is, however, hampered by the presence of preadsorbed oxygen.[238]

Nitromethane adsorbed on Ni{111} surfaces decomposes at 370 K to form mainly HCN, H_2, and adsorbed oxygen. Adsorption at slightly above room temperature results in an enhancement of the C–H stretching vibration bands of adsorbed nitromethane. The enhancement is attributed to the alignment of dipoles of adsorbed nitromethane molecules.[190] Pyrrole is adsorbed on Ni{100} surfaces with its molecular plane parallel to the surface. The pathway of thermal decomposition of adsorbed pyrrole changes depending upon the heating rates employed. The change is explained by assuming a change from a decomposition reaction to a polymerization reaction with increasing heating rates.[239] Similar behavior was observed in the adsorption and decomposition of thiophene on Ni{111}.[240] The adsorption of methyl-substituted benzenes on Ni{100} was observed with IR spectroscopy in experiments designed to determine the effect of methyl substitution on the interaction of the benzene ring with the surface. All molecules are adsorbed with the ring parallel to the surface at the initial stage of adsorption. Methyl substitution causes a decrease in the binding energy of the ring to the surface. The decrease remains unchanged regardless of the number of substituents, but decomposition products in the reaction with the Ni surface change depending upon the placement of the methyl groups. The experimental results were discussed and compared with extended Hückel calculations.[241]

The polarization modulation method was utilized in combination with a tunable diode laser as the IR source for the observation of oxygen adsorption on an aluminum substrate.[33] Adsorbed oxygen shows features at 890, 770, and 580 cm^{-1} at submonolayer coverage. The highest frequency feature is ascribed to the vibration of oxygen that constitutes portions of an Al_2O_3-like structure, and the latter two features are ascribed to perturbed subsurface and surface Al–O vibrations, respectively.

2.1.4. Infrared Ellipsometric Spectroscopy

Ellipsometry and ellipsometric spectroscopy (ellipsometric measurement as a function of wavelength) using near-ultraviolet and visible light have been widely utilized for the investigation of solid surfaces and thin films between two phases.[242] While they are employed for investigating thin films on solid substrates in combination with IR reflection spectroscopy,[243] ellipsometric spectroscopy has been extended to the infrared region by Dignam *et al.*[244]

Infrared ellipsometric spectroscopy (IRES)[245,246] is a spectroscopic technique similar to polarization modulation IR spectroscopy. Both of these techniques are feasible for the *in situ* observation of solid surfaces in absorbing media over a wide range of temperatures, but the former provides more extensive information than the latter. Since, in IRES, relative phase shifts are measured over an extended wavelength region in addition to amplitude ratios, the absorption coefficient of the surface film can be calculated as a function of wavelength, giving an absorption spectrum without distortion.[247] Furthermore, changes in the optical properties of the surface layer of the solid substrate caused by the adsorption can be determined through the measurement of IRES, i.e., the change in the properties of the conduction electrons in the surface layer gives rise to background shifts in the absorption and dispersion spectra obtained with IRES.[245] In IRES, however, measurements must be repeated upon changing the passing axis of either of the two stationary plane polarizers, causing the prolongation of the measurement time.[247,248] The use of a polarization modulation ellipsometer reduces the measurement time, but the modulation instruments require more stationary polarizers than in polarization modulation reflection spectroscopy, resulting in a sizable loss of infrared energy, which may reduce the S/N ratio of the obtained spectra.

The Fresnel reflection coefficients, r_p and r_s, of a planar interface for p- and s-polarized light can be written, respectively, as

$$r_p = |r_p| \exp(i\theta_p), \ r_s = |r_s| \exp(i\theta_s) \tag{18}$$

where $|r_p|$ and $|r_s|$ represent the ratio of amplitude of the reflected to the incident light, while θ_p and θ_s represent the phase shift upon reflection. In ellipsometry, the ratio χ of the complex reflection coefficients is determined by

$$r_p/r_s = \chi = |\chi| \exp\{i(\theta_p - \theta_s)\} \tag{19}$$

where $|\chi| = |r_p|/|r_s|$. Equation (19) is generally written using conventional parameters ψ and Δ as

$$\chi = \tan\psi \exp(i\Delta) \tag{20}$$

Stobie *et al.* defined a quantity, L, in terms of the natural logarithms of the ratio of the complex reflection coefficients:[249]

$$L = -\{\ln(\tan\psi) + i\Delta\} \tag{21}$$

They also defined the ellipsometric reflectance absorbance, A_e, using common logarithms, as

$$A_e = \log[|r_p/r_s|_0^2/|r_p/r_s|^2] \tag{22}$$

where the subscript "$_0$" identifies the quantity connected with the clean surface. With $\Delta L = L - L_0$, the real part of ΔL is related to A_e by

$$A_e = (2/2.303)Re(\Delta L) \tag{23}$$

Since $|r_s|^2/|r_s|_0^2 \simeq 1$ when the film-covered metal is highly reflective, A_e determined as a function of wavenumber gives the elliposometric absorbance spectrum of the thin film. The imaginary part of ΔL gives the dispersion spectrum of the film.[250]

Strictly speaking, however, there exists the contribution to the measured absorbance spectrum from absorbing gas-phase molecules present in the interphase whose effective thickness is $\sim\lambda_{vac}/\sqrt{n_3^2 + k_3^2}$.[245] This is also the case in the observation of polarization modulation IR reflection spectra, as described in the preceding section. Further attention must be paid to the fact that the ellipsometric absorbance spectrum is not the spectrum of the absorption coefficient of the surface film and is not free from distortion.

Both dispersive spectrometers and interferometers have been used in IRES measurements. An ellipsometric unit, in which two stationary plane polarizers are placed immediately before and after the reflecting sample, is generally set between the monochromator (or interferometer) and the detector. In the polarization modulation ellipsometric measurement, a rotating polarizer or a photoelastic modulator is introduced next to the first stationary polarizer. An example of the optical arrangement is shown in Fig. 43.[251] The photoelastic modulator is conveniently used in combination with the interferometer by virtue of its high modulation frequency, but has a disadvantage that the exact half-wave retardation is accomplished at only one wavenumber for a fixed power applied to the piezoelectric driver, as mentioned already. This problem can be removed by the use of a polarizing Michelson interferometer, in which a metal-grid polarizing beam splitter is mounted in place of the conventional beam splitter, and the optimum conditions can then be obtained throughout the mid-infrared region.[252,253]

Infrared ellipsometric spectroscopy was applied to the investigation of CO adsorption on Ni{110}, over a wide pressure and temperature range,[254] and the catalytic methanation reaction on Ni{110}.[255] In the investigation of methanol decomposition on Ni{100}, the IR data imply the existence of two intermediates of methoxy and formyl (–CHO) moieties.[251]

Blayo and Drévillon have observed *in situ* the early stage growth of

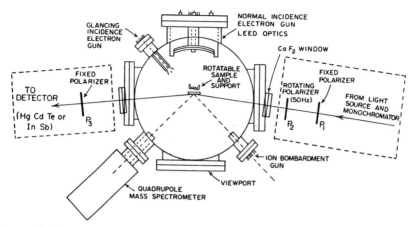

Figure 43. Schematic diagram of the sample section of a polarization modulation ellipsometer used for the measurement of clean surfaces [Reprinted from F. L. Baudais, A. J. Borschke, J. D. Fedyk, and M. J. Dignam, *Surf. Sci.* **100**, 210 (1980)].

a–Si:H and microcrystalline silicon (μc–Si) films with infrared phase modulated ellipsometry.[256] The presence of a hydrogen-rich overlayer has been detected for a–Si:H films growing on glass substrates by plasma enhanced chemical vapor deposition.[257] Plasma deposited μc–Si films grow inhomogeneously, and the hydrogen plasma etching of the films causes the removal of the hydrogen-rich overlayer and induces the transition of the film structure from amorphous to microcrystalline.[258]

2.2. Wavelength Modulation Method

The wavelength modulation method involves the observation of the wavelength derivative of the reflection spectrum.[11,259] The observation of the first derivative of the normal spectrum is quite effective for the detection of weak but sharp features superimposed on a smooth and large background and is often used in surface spectroscopies such as Auger electron spectroscopy and some other electron excited spectroscopies.

The wavelength modulation can be attained by vibrating either the monochromator entrance slit,[260] the mirror adjacent to the entrance slit,[261] or a sodium chloride disc movable about an axis parallel to the exit slit.[262] The sensitivity attained by the wavelength modulation is higher than that of the polarization modulation as long as sharp features are involved. An example of spectra obtained with the wavelength modulation method is shown in Fig. 44.

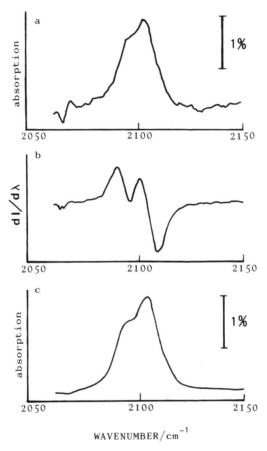

Figure 44. Conventional reflection and wavelength modulated spectra of CO adsorbed on Cu{311} (a) conventional, (b) derivative, (c) integrated derivative [Reprinted from J. Pritchard, T. Catterick, and R. K. Gupta, *Surf. Sci.* **53**, 1 (1975)].

The numerical integration of the derivative spectrum (spectrum b) yields the very smooth spectrum shown in this figure, because random noise signals nearly cancel each other out as a result of the integration. Ryberg, on the other hand, recorded the second derivative of the reflectance as a means of avoiding the errors introduced in the process of integration.[263]

An important disadvantage of the wavelength modulation method is its high sensitivity to absorbing gas-phase species, which necessitates the evacuation of the entire spectrometer to eliminate atmospheric absorptions. *In situ* observation of solid surfaces in absorbing media is still difficult with this modulation method, even if the evacuated spectrometer is used. Another disadvantage is its low sensitivity to broad spectroscopic features of surface species. It is not widely used at present because of these disadvantages.

Pritchard and his collaborators observed the adsorption of CO on various faces of copper single crystals with the wavelength modulation method and investigated the mutual interaction of adsorbed CO.[264] They also studied the influence of defect sites on the IR spectra of CO adsorbed on a surface vicinal to Cu{110},[265] and attributed the high frequency component of a doublet C–O stretching band to CO molecules adsorbed at the defect sites.

Perdeuteroethane, C_2D_6, adsorbed on a clean Cu{110} surface at 77 K exhibits a feature at 2210 cm^{-1} in the wavelength modulated spectrum. This band can be assigned to the degenerate C–D stretching vibration accompanying a dynamic dipole perpendicular to the C–C axis. Gas-phase C_2D_6 has parallel bands at 2087 and 2111 cm^{-1}. Although these two bands have intensity approaching that of the degenerate vibration, they are missing in the modulated spectra of adsorbed C_2D_6. This implies that the ethane molecule is adsorbed with its C–C bond parallel to the surface.[261]

The adsorption of CO on Pt{111} has been investigated with wavelength-modulated IR spectroscopy. The high resolution of IR reflection spectroscopy revealed two adsorption states in the bridging region, corresponding to two-fold bridge and three-fold hollow sites.[266] The adsorption of CO on a stepped platinum surface has been investigated using the wavelength modulation method.[134] The results obtained in this investigation were described in Section 1.4.2 of this chapter.

Wavelength-modulated IR spectroscopy combined with temperature programmed desorption has been used for the investigation of the adsorption and decomposition of formic acid on Cu{110}. Acid monomer was found on the copper surface at 120 K. The bridging formate forms on the surface via deprotonation at ~270 K. A combination vibration band of adsorbed species was first observed in this work at 2950 cm^{-1}. This band has an intensity comparable to that of the closely located C–H stretching band by virtue of a Fermi resonance.[267]

NO is adsorbed on Pt{111} forming at first bridge-bonded and subsequently linearly bound species at 95 K. The linearly bound NO saturates at a low coverage $\theta = 0.25$. The coverage-dependent frequency shift of the two species was also investigated.[268] Methoxide is formed on oxidized Cu{100} when methanol is introduced to the atomic oxygen-covered copper surface. The orientation of the methoxy group tilted to the surface has been determined from the measurement of symmetric and asymmetric stretching modes of the CH_3 group.[269] The wavelength modulation method was employed also in the observation of the low frequency metal–CO stretching vibration.[270]

3. Potential Modulation Method

3.1. Basic Principles and Instrumentation

The change in electrode potential causes a change in concentration of species present in the electrode/electrolyte interphase and those adsorbed on the electrode surface. In consequence, if the electrode potential oscillates between two fixed potentials, corresponding to two distinct states of the electrode surface, and the modulated (AC) component of the IR light reflected on the electrode surface is demodulated, information on the mechanism of the electrochemical reaction and adsorption on the electrode surface can be obtained without interference due to the signals of species in the bulk solution.

This modulation technique, called the electrochemical modulation method, was first employed in the near-ultraviolet and visible region, where the most widely used solvent, water, is transparent; numerous works have investigated passive films, the mechanism of electrochemical reaction, adsorption on electrodes, and so on.[271–273] However, a number of compounds have no absorption in the near-ultraviolet and visible region, and the observation of electronic states does not provide detailed information about the structure of surface species. The desire to observe *in situ* vibrational spectra of species on electrode surfaces has lead to the combination of the electrochemical modulation technique with resonance (or surface-enhanced) Raman spectroscopy, in which visible light is used as the probe.[274,275]

The potential modulation measurement, in which no electrochemical process is involved, is sometimes called the field effect modulation method and has been used for many years in the infrared region for the observation of absorptions due to free holes and electrons present in the space charge region and surface states of semiconductors,[276,277] as well as for investigating the field-induced orientation of liquid crystal molecules on and near an electrode surface.[278]

The extension of the electrochemical modulation method to the infrared region, however, was achieved years later than the field effect modulation, because infrared light is absorbed by water to a large extent.[279,280] This difficulty has been overcome by a group of scientists at the University of Southampton in England.

They calculated that the infrared radiation reflected from metal electrodes in aqueous solutions retains sufficient strength for the observation of spectra even in the region where the absorption due to the O–H stretching vibration of water is present, provided the thickness of the solution is less than ~1 μm. By modulating the electrode potential, then, they first observed infrared spectra of indole adsorbed on a Pt electrode in acetonitrile. They also obtained spectra of species on a Pt electrode surface in sulfuric acid solution.[281]

The potential modulation is performed by superimposing a square wave AC signal of ~10 Hz from a function generator onto a fixed bias voltage applied to the working electrode by a potentiostat. The signals of species in the electrode/electrolyte interphase and those adsorbed on the electrode change in accordance with the potential modulation, provided the electrochemical change of the surface species is reversible and sufficiently fast. The AC component of the detector output is processed by a lock-in amplifier tuned to the AC frequency to give the spectrum of the surface species. The signals of the solvent and solution species immune to the potential modulation are removed in the demodulation process.[279,280] The signals from surface species immune to the potential change are also removed from the spectrum obtained by the electrochemical modulation method in contrast to the polarization modulation method, in which all the absorbing surface species are detected as long as their vibrational modes accompany dynamic dipoles normal to the metal surface. In addition, the low modulation frequency in electrochemically modulated IR reflection spectroscopy (EMIRS) makes the rapid acquisition of a spectrum difficult, because the modulation frequency must be sufficiently high compared with the wavenumber scanning rate. It should be noted that the difference between two spectra of the surface species in two distinct states is recorded and the spectrum of the surface species in each state remains unknown in EMIRS. It follows that the quantitative measurement of surface species and the determination of band peak position are difficult with the electrochemical modulation method. Beden *et al.* proposed an integration procedure to determine absorption peaks as well as to reduce the noise of the spectra[282] and applied this technique to the investigation of the adsorption of methanol on polycrystalline Pt and Pt{111} surfaces.[283]

The polarization modulation method has been coupled successfully with interferometers; however, the electrochemical modulation method is generally used in combination with dispersive spectrometers, because the frequency of potential modulation should be ~10 Hz or less as a result of the use of thin solution layer cells. Interferometers may be used in the measurement of electrochemical modulation if the electrode material is a semiconductor. The measurement at electrode surfaces, then, can be performed with the internal reflection method, which allows the use of conventional electrolytic cell. A relatively high modulation frequency of about 1 kHz can be used with the internal reflection measurement. This frequency makes the combination of the electrochemical modulation method with interferometers possible, provided the interferometer is operating at a very low modulation frequency.[284,285] Ozanam and Chazalviel successfully combined the electrochemical modulation technique with an interferometer mounted with a very low-moving mirror at a speed 1–10 μm/s. This combination has been applied to the *in situ* observation of species near and on a semiconductor electrode in the study of diffusion-limited and multistep electrochemical reactions. Although this technique requires long measurement times, a sensitivity of ~10^{-6} in $\Delta I/I_0$ can be achieved with the single-reflection arrangement. The modulation frequency can be changed from a few kHz down to some 10 Hz using this technique. This is advantageous in the investigation of the mechanism of electrochemical reactions.[286]

Interferometers are, on the other hand, widely utilized in subtractively normalized interfacial FTIR spectroscopy (SNIFTIRS), where the spectrum recorded at a reference potential is subtracted from that recorded at the potential of investigation.[279]

The thickness of the solution layer in electrolytic cells employed in EMIRS and SNIFTIRS must be very thin, as already mentioned. An example of thin solution layer cells is shown in Fig. 45.[287] The cell shown in Fig. 39 may be used in EMIRS and SNIFTIRS. Silicon, CaF_2, As_2Se_3, IRTRAN 4 (ZnSe), and KRS 5 (TlBr-TlI) are used for windows, because IR-transparent crystals such as KBr and NaCl generally used in IR measurements are soluble in water. Silicon and CaF_2 are most frequently used as windows of the electrolytic cells. Since CaF_2 is transparent in the visible region, its use facilitates the visual inspection of the electrode. However, it is opaque below 1000 cm^{-1}. In addition, windows of CaF_2 become blurred after prolonged use, because they are slightly soluble in water.

High purity silicon is transparent in the mid- and far-infrared regions, and it is insoluble in water as well as in dilute acid solutions. Since it is opaque, however, to the visible light, the electrode surface cannot be observed during cell alignment and measurement. In addition, the refractive index of silicon is high, giving rise to high reflection loss of IR light. For this reason, the arrangement shown in Fig. 45 is used. In this arrangement, the infrared beam is incident at the Brewster angle ($\phi_B = 74°$ for silicon) on the window (W), because *p*-polarized light, which interacts with species on the metal surface, is totally refracted into the window upon incidence at the Brewster angle. However, the IR beam is incident on the silicon/solution interface at an angle different from the Brewster angle (the pseudo-Brewster angle for absorbing solution), and it

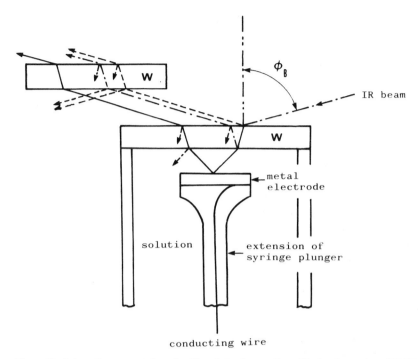

Figure 45. Schematic representation of a thin solution layer cell used in the observation of EMIR and SNIFTIR spectra. W, window material; ——, mostly *p*-polarized light; – –, *s*-polarized light; —·—·—, *s*- and *p*-polarized light.

is often not collimated. The intensity of *p*-polarized light, therefore, is sizably reduced during the transit through the cell. The intensity of *s*-polarized light, which is of no use for the measurement of surface species, is remarkably attenuated, because the reflectance of *s*-polarized light is about 70% at the Brewster angle. Its intensity can be further attenuated by placing a silicon plate having parallel faces parallel to the window surface, as shown in Fig. 45.

The material IRTRAN 4 is transparent above 650 cm⁻¹ and partially transparent to visible light, but deteriorates in contact with acid solutions. Deeply orange-red colored KRS 5 is transparent down to 250 cm⁻¹, but is poisonous and more soluble than CaF_2 in water. In addition, its relatively high refractive index produces reflection losses of infrared light. Brown colored As_2Se_3 is transparent above 900 cm⁻¹ and insoluble in cold water but is soluble in alkaline solutions and has a relatively high refractive index.

The working electrode is sealed into or glued to a glass tube attached to the glass syringe plunger. The electrode can be pushed against the window of the cell to obtain a thin layer of solution for adequate transmission of IR light. For the exposure of the electrode to the bulk solution, the electrode is withdrawn into the body of the cell.[279]

Attention must be paid to the use of thin solution layer cells because of the awkward consequences they may cause. The first is the slow diffusion of reaction products from the electrode and that of reactant species to the electrode, which makes high frequency modulation inapplicable and may induce unexpectable changes of pH near the electrode surface. The second is the difference between the potential applied to the working electrode and the actual potential of the electrode. This difference may be appreciable when a dilute electrolyte solution is used or when the electrode process demands an electric current of high density. There exist other problems, which are described in Section 2.1.3.2 of this chapter. In spite of these problems, the electrochemical modulation method is widely employed in the investigation of electrode processes, because weak signals of the order of 10^{-6} in $\Delta R/R_0$ can be detected with this modulation method in the wavenumber region, where sufficient transmission of IR radiation is obtained.

In the investigation of semiconductor electrode surfaces, the electrochemical modulation measurement can be carried out with the internal reflection arrangement using conventional electrolytic cells. The modulation of the potential applied to semiconductor electrodes induces, however, the electron transition to or from the space-charge layer and surface states in addition to the electrochemical change of species in the interphase. The free-carrier absorption produces broad bands, and the transitions involving surface states yield somewhat broad or sharp bands. These bands have been the subject of additional research.[288,289] In the investigation of electrochemical processes on semiconductor electrodes, the broad band, which is superimposed over the signals from the surface species, is approximated to a theoretical ("best-fit") background and subtracted from the total spectrum to get the spectrum of the surface species.

3.2. Applications

3.2.1. Electrochemically Modulated IR Reflection Spectroscopy (EMIRS)

Because methanol is relevant to fuel cells, the mechanism of its oxidation on Pt electrodes has been investigated for many years with electrochemical techniques, but the reaction intermediates have been the subject of controversy. The methoxy group, COH, CHO, and CO have been claimed as the major surface species formed during the electro-oxidation of methanol.[290–292] Electro-chemically modulated IR reflection spectroscopy has played an important role in the resolution of the controversy.

In situ observation of EMIR spectra of a Pt electrode surface in contact with an aqueous methanol solution has given the spectrum shown in Fig. 46.[293] This figure shows a positive-going band at 1850 cm⁻¹, which corresponds to an increase in the concentration of species associated with the band upon making a cathodic shift of electrode potential from 0.45 to 0.05 V (NHE). The strong derivative-type feature having a peak at 2051 cm⁻¹ and the trough at 2085 cm⁻¹ stem from the shift of a band to a higher frequency with the anodic sweep of electrode potential from 0.05 to 0.45 V. The wavenumber of these two bands suggests the presence of bridge-bonded and terminally bound CO on the Pt electrode. The nearly symmetric shape of the bridge-bound CO band at 1850 cm⁻¹ probably corresponds to the disappearance of this species at 0.45 V. The results indicate the presence of CO on the Pt electrode as a dominant adsorbate.

This EMIR observation demonstrated that terminally bound CO is the main adsorbate on smooth polycrystalline Pt electrodes during the electro-oxidation of methanol in aqueous acidic solutions containing a relatively high concentration of methanol, and the adsorbed CO is generally considered to be

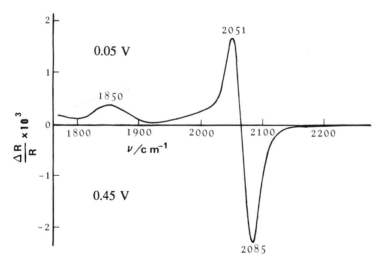

Figure 46. EMIR spectrum from Pt electrode/aqueous methanol-sulfuric acid solution interphase. Electrode potential = 0.25 ± 0.20 V (NHE), modulation frequency = 8.5 Hz, concentration of methanol = 0.5 mol/l, scanning rate = 0.0127 µms⁻¹ [Reprinted from B. Beden, C. Lamy, A. Bewick, and K. Kunimatsu, *J. Electroanal. Chem.* **121**, 343 (1981)].

the poisoning species in the oxidation reaction. However, the observation of spectra in a wide wavenumber region has shown that a change in experimental conditions brings about a marked change in adsorbed species on the electrode surface.

The species adsorbed on Pt electrodes during the electro-oxidation of methanol change depending upon the crystal face. Whereas terminally bound CO is the predominant adsorbate on the Pt{110} surface, as on the polycrystalline Pt surfaces, bridged CO and formyl-like species are adsorbed on Pt{100} and Pt{111} surfaces in addition to terminal CO in solutions containing methanol at a high concentration. Furthermore, the adsorption of CO is greatly reduced if the electrode surface is roughened. The occupation of active sites by terminal CO, therefore, is not the sole cause of the poisoning of Pt electrodes in the electro-oxidation of methanol in acidic solutions.[294]

The EMIR observation in a wide wavenumber region from 1400 to 2700 cm^{-1} has revealed that, in addition to terminally bound CO, formate species are present on polycrystalline Pt and Pt{111} as an important adsorbate in acidic solutions containing methanol at a relatively high concentration (0.1 mol/l). The decrease in concentration to 10^{-3} mol/l results in a drastic decrease in the intensity of the CO band. In addition, the signals assignable to methyl formate appeared in spectra from the Pt{111} surface in the initial stage of adsorption.[283]

Species adsorbed on Pt electrodes change remarkably in changing the medium from acidic to alkaline. Only a small amount of adsorbed CO resulted from the dissociative adsorption of methanol on a Pt{100} electrode in carbonate solutions, even when the concentration of methanol was relatively high. The electrode surface maintains its catalytic activity for a long time in carbonate solutions.[295]

Solomun investigated poisoning species on Pt, Rh, and Pd electrodes during the electro-oxidation of formaldehyde in acidic solutions by means of EMIRS.[296] The IR spectra obtained show the presence of CO adsorbed at various sites on the electrode surfaces. The stability of adsorbed CO even at extreme negative potentials and the absence of COH, another presumed poisoning species, on the electrodes imply that the adsorbed CO species is responsible for the poisoning of Pd, Pt, and Rh electrodes in the oxidation of formaldehyde in acidic solutions.

The adsorption of CO has been studied using Pt, Rh, Pd, and Au as the electrode.[297-299] Both bridge-bound and terminally bound CO were detected on all of these electrodes. The potential and coverage dependence of the adsorbed CO, as well as the oxidation of CO into CO_2, were studied. The adsorption of CO on noble metal electrodes with or without modifying foreign metal atoms has been reviewed, comparing the experimental results of EMIRS and conventional electrochemical measurements with extended Hückel calculations.[300]

Electrochemically modulated IR reflection spectroscopy has been employed in the investigation of the structural change of iridium oxide during the process of electrochromism. A feature due to the O–H stretching vibration was found in the bleached oxide film. This feature is attributed to IrOOH species produced by the injection of protons and electrons into the oxide.[301]

Electrochemically modulated IR reflection spectroscopy in the internal reflection arrangement combined with the transmission measurement has been applied to the investigation of porous silicon layer formation on a low-doped Si surface in contact with hydrofluoric acid solution. The IR spectra exhibit a broad SiH band around 2100 cm^{-1}. This band splits into three sharp features at 2085, 2115, and 2140 cm^{-1} after withdrawal of the electrode from solution followed by the evaporation of solvent. The broadening of the feature, therefore, is ascribed to the solvent effect. The three sharp features appear also in the spectra of an n–Si/HF solution interface under cathodic potentials, at which the surface SiH groups are in contact with hydrogen bubbles present at the interface. A very short lifetime has been deduced for the intermediate species in the process of anodic dissolution of Si in HF.[302]

The modification of semiconductor and metal electrodes with heteropolyacid anions brings about the enhancement of the kinetic parameters of the hydrogen evolution reaction arising from the increase in charge transfer across the electrode/electrolyte interface. The reductive modification process of an n-Ge electrode with silico- and phosphotungstate ion, and the species formed on the modified electrode surface, have been investigated with EMIRS and the measurement of cyclic voltammograms.[303] The spectra of intermediates formed in the modification reaction show the weakening of W–O bonds of the adsorbed species and the strong interaction of water with the $W_{12}O_{36}$ skeleton of the adsorbate. Heavily hydrated heteropolyacid anions are present on the modified Ge surfaces. The presence of penetrated hydrogen in the Ge electrode is also found in the spectra of the modified electrode.

3.2.2. Subtractively Normalized Interfacial FTIR Spectroscopy (SNIFTIRS)

The combination of interferometers with electrochemical modulation is difficult in the external reflection measurement, as mentioned above. The synchronization of the modulation of the interferometer with that of the electrode potential is not easy. Subtractively normalized interfacial FTIR spectroscopy, then, was developed for the observation of electrode processes in the external reflection arrangement with FTIR spectrometers.[279]

For the observation of reversible electrode processes, several techniques can be used in SNIFTIRS.[279] For the measurement of reaction intermediates or metastable species, a small number of interferograms are acquired and co-added with a fast scanning FTIR spectrometer at the reference electrode potential, e.g., at a potential at which no adsorption of species under study occurs. The potential is then shifted to a desired value for acquiring interferograms of the sample spectrum. This procedure is repeated to improve the S/N ratio. The SNIFTIR spectrum is obtained by subtracting the reference spectrum from the sample spectrum. This measurement may be considered a potential modulation method at an extremely low frequency, but it is called generally SNIFTIRS or potential difference IR spectroscopy (PDIRS).[304,305] Double-beam FTIR spectrometers may be employed for SNIFTIR observations using two identical cells maintained at the reference and the working potentials, respectively.

When the electrode process is irreversible, a series of interferograms is co-added at the reference potential to enhance the S/N ratio. The electrode potential is then changed to a desired value for measuring the sample spectrum, and the interferograms are co-added at that potential. The difference between the sample spectrum and the reference is then recorded in this single-step technique. However, drifts in the measurement system may produce serious problems in prolonged measurements. The repetition of the acquisition of a small number of interferograms at the reference and the sample potentials followed by the exchange of electrolyte solution may provide good results.[304]

The behavior of cyanide ions on electrodes is of practical as well as fundamental interest. Cyanide ion is widely used as an important additive in the electroplating of metals and is suitable for the investigation of adsorption on metal electrodes, because a wide double-layer region where no electrochemical reaction occurs is available in pseudohalide ion/metal electrode systems. Hence, the potential dependence of the C–N stretching vibration of cyanide ions adsorbed on a smooth polycrystalline palladium surface was investigated in perchlorate solutions.[306,307]

Cyanide ions adsorbed on Pd electrodes exhibit two features at about 2060 and 1980 cm^{-1}, assignable to terminally and bridge-bound CN^-, respectively. These two features shift to higher frequencies when the electrode potential is made more positive, but the shift is larger for the bridge-bound species than for the terminal species.[307] This contrasts with the CO–Pd system where the peak frequencies of the terminally and bridged CO species show nearly the same potential dependence attributable to the electrochemical Stark effect.[298] The frequency shift of the CN^- band has been attributed to the change in back-donation from the metal to the adsorbates rather than the electrochemical Stark effect. The spectroscopic features of surface precursors to solution species of palladium cyanide complexes have also been detected.[307]

The adsorption of another pseudohalide ion, SCN^-, on gold in an aqueous solution was investigated with SNIFTIRS and surface enhanced Raman spectroscopy.[308] Adsorbed SCN^- ions showed a C–N stretching band around 2125 cm^{-1}, attributable to a S-bound adsorbate. The peak frequency and its potential dependence changed only slightly when the smooth surface of the electrode was electrochemically roughened. The measurement of the Raman spectrum of adsorbed SCN^- ion on a roughened Au electrode gave nearly the same results. This implies that the site active for the surface-enhanced Raman scattering does not differ in the main from the site detected by SNIFTIRS.

The adsorption and electro-oxidation of benzoate and cyanate anions were studied using Au and Pt electrodes in an aqueous perchloric acid solution. In addition to the repeated change of potential between the reference and the sample values, the single step technique was used for the observation of irreversible processes.[305] Benzoate ion is adsorbed reversibly through the carboxylate group on Au electrodes with the plane of the benzene ring normal to the electrode surface. The single step observation has revealed the existence of irreversible adsorption of the ion on Pt electrodes. The benzoate ion adsorbed on Pt electrodes converts into CO_2 at potentials corresponding to the beginning of surface oxide formation. Cyanate ion adsorbed on Pt and Au electrodes is oxidized at potentials somewhat less positive than those of surface oxide formation. At the same time, the cyanate ions in solution migrate to the electrode surface to be adsorbed or converted into isocyanic acid.

Several measurements of SNIFTIR spectra have been carried out using nonaqueous solutions. Since the absorptions of nonaqueous solvents are generally weaker than those of water, a solution layer as thick as 50 μm may be used in the SNIFTIR observation.

Tetracyanoethylene has very interesting properties. It is a typical molecule acting as a good electron acceptor to organic donors and metals. In addition, its anions exhibit, in the solid state, infrared bands assigned to totally symmetric modes, which should be infrared-inactive, as a result of the vibronic coupling of charge transfer. The behavior of tetracyanoethylene anions on a Pt electrode has been observed by SNIFTIRS in acetonitrile solutions. The results obtained imply that the tetracyanoethylene anion radical generated electrochemically is adsorbed on the electrode with its molecular plane parallel to the electrode surface.[309] Detailed information about the adsorbed anions is obscured by the strong signals from solution species and the solvent, but the use of the Kretschmann ATR arrangement greatly reduces this difficulty. An infrared band assignable to a totally symmetric mode was observed in this arrangement.[310]

The adsorption of hydrogen on iron electrodes was investigated by SNIF-TIRS during the hydrogen evolution reaction.[311] Features arising from the adsorbed hydrogen were located at 2060 (1485 in D_2O) and 980 cm^{-1} and assigned to Fe–H stretching and bending vibrations, respectively. The adsorption of borate ions on passivated iron was investigated with relevance to the stability of the passive film on iron electrodes by means of the *in situ* observation of SNIFTIR spectra in borate buffer solutions.[312] The results obtained suggest that borate ions are strongly adsorbed on the surface, giving rise to a decrease of the diffusion of Fe^{2+} ions through the film. Adsorbed borate ions thus reduce the dissociation of iron. When sulfate ions are coexistent with borate, however, the former ions displace the adsorbed borate ions, causing the deterioration of the passive film.

Subtractively normalized interfacial FTIR spectroscopy was also applied to the observation of semiconductor/electrolyte interfaces for investigating the adsorption of ions and neutral molecules on the semiconductor electrode from aqueous and nonaqueous solutions.[214] Thiorea is adsorbed on *p*-silicon electrodes through the sulphur atom in aqueous solutions. A negative shift of the electrode potential results in an increase in quantity of adsorbed thiourea. In a solution containing sodium acetate, neutral acetic acid molecules are adsorbed on *p*-silicon electrodes, and the adsorption reaches a maximum at a potential approximating the potential of zero charge.

In acetonitrile solutions containing sodium acetate, the species adsorbed on *p*-GaP electrodes vary depending upon the electrode potential. At cathodic potentials, acetonitrile is adsorbed strongly on the electrode, but acetate ion begins to be adsorbed when the electrode potential approaches the potential of zero charge, and this effect increases, replacing adsorbed acetonitrile upon further anodic shift of the potential. Ammonium ion is adsorbed on gallium phosphide electrodes from acetonitrile solutions via two hydrogen atoms.

3.2.3. Field Effect Modulation Spectroscopy

The potential modulation method has also been applied to systems in which no electrochemical process is involved. The modulation method has been employed in the study of the dynamic behavior of liquid crystal molecules under an applied electric field,[278] in addition to the investigation of the electronic states of semiconductor surfaces.[277]

The dynamic behavior of nematic 4-*n*-pentyl-4′-cyanobiphenyl (5CB) liquid crystal molecules has been investigated through the measurement of the potential modulation IR ATR spectra of a thin liquid crystal layer.[278] A thin layer of 5CB of about 20 μm in thickness was formed between two electrodes of a Ge hemicylinder prism and an indium–tin oxide (ITO) conducting glass plate. The ITO plate surface was rubbed in one direction, coinciding in the ATR cell with that of the oscillating electric field of the incident s-polarized IR light.

The alignment of the liquid crystal molecules has been determined by the observation of polarization modulation spectra at various incident angles. The spectra show that 5CB molecules are aligned with their long axes parallel to the rubbing direction in the close vicinity of the germanium electrode surface as well as at the midpoint of the layer in the absence of an external electric field. The application of a DC field causes a change in alignment of the liquid crystal molecules in the bulk layer, but the change is reduced remarkably in the vicinity of the germanium surface.

The field effect modulation measurement was carried out in the C–N stretching region, because the dynamic dipole of the vibration is oriented nearly parallel to the direction of the long axis of the liquid crystal molecule. The absorption signals measured as a function of the amplitude of the AC modulation field are shown in Fig. 47. The upward and downward features represent the increase and decrease in absorption intensity upon application of the electric field, respectively.

This figure shows that, whereas the absorption of p-polarized light increases upon application of an electric field, that of s-polarized light decreases.

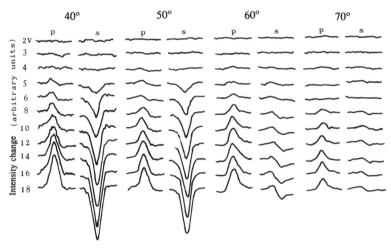

Figure 47. Field effect modulation spectra of a 5CB liquid crystal layer as a function of applied AC voltage. Angle of incidence is indicated at the top. Modulation frequency = 10 Hz [Reprinted from A. Hatta, H. Amano, and W. Suëtaka, *Vib. Spectrosc.* **1**, 371 (1991)].

This indicates that the applied field induces a change in alignment of the liquid crystal molecules because of a positive dielectric anisotropy of the molecule.[313] The intensity change decreases with increasing angle of incidence. As the penetration depth of the evanescent wave decreases with the increase of incident angle, the observed decrease in signal strength indicates that the molecules present in the close vicinity of the germanium electrode surface react only slightly to the applied AC field. No signal appears in the spectra when the applied electric field is weaker than 4 V, showing that the presence of the threshold field strength for the induced alignment is at about 3.5 V.

An increase in modulation frequency results in the decrease of the signal strength of the C–N stretching band. This shows that the liquid crystal molecules cannot follow the rapid change in the electric field. The threshold field strength also increases with an increase in modulation frequency.[278]

References

1. R. P. Eischens, W. A. Pliskin, and S. A. Francis, *J. Chem. Phys.* **22**, 1786 (1954).
2. A. N. Terenin, *Zh. Fiz. Khim.* **14**, 1362 (1940).
3. L. H. Little, *Infrared Spectra of Adsorbed Species,* Academic, London (1966).
4. S. A. Francis and A. H. Ellison, *J. Opt. Soc. Am.* **49**, 131 (1959).
5. L. S. Bartell and D. Churchill, *J. Phys. Chem.* **65**, 2242 (1961).
6. J. Leja, L. H. Little, and G. W. Poling, *Trans. Inst. Min. Metall.* **72**, 407 (1963).
7. R. W. Hannah, *Appl. Spectrosc.* **17**, 23 (1963).
8. R. G. Greenler, *J. Chem. Phys.* **44**, 310 (1966).
9. W. Suëtaka, *Jasco. Rept.* **4**, 1 (1967).
10. Y. J. Chabal, *Surf. Sci. Rept.* **8**, 211 (1988).
11. P. Hollins and J. Pritchard, in *Vibrational Spectroscopy of Adsorbates* (R. F. Willis, ed.), Springer, Berlin (1980) p. 125.
12. H. A. Pearce and N. Sheppard, *Surf. Sci.* **59**, 205 (1976).
13. R. G. Greenler, D. R. Snider, D. Witt, and R. S. Sorbello, *Surf. Sci.* **118**, 415 (1982).
14. J. D. E. McIntyre and D. E. Aspnes, *Surf. Sci.* **24**, 417 (1971).
15. O. S. Heavens, *Optical Properties of Thin Solid Films,* Dover, New York (1965).
16. D. M. Kolb and J. D. E. McIntyre, *Surf. Sci.* **28**, 321 (1971).
17. N. Morito and W. Suëtaka, *J. Jpn. Inst. Metals* **36**, 1131 (1972).
18. H. Pfnür, D. Menzel, F. M. Hoffman, A. Ortega, and A. M. Bradshaw, *Surf. Sci.* **93**, 431 (1980).
19. V. M. Bermudez, *J. Vac. Sci. Technol.* **A10**, 152 (1992).
20. A. Hatta, H. Matsumoto, and W. Suëtaka, *Chem. Lett.* **1983**, 1077.
21. A. Hatta, T. Wadayama, and W. Suëtaka, *Anal. Sci.* **1**, 403 (1985).
22. Y. Ishino and M. Ishida, *Appl. Spectrosc.* **46**, 504 (1992).
23. N. Sheppard, R. G. Greenler, and P. R. Griffiths, *Appl. Spectrosc.* **31**, 448 (1977); D. H. Chenery and N. Sheppard, *Appl. Spectrosc.* **32**, 79 (1978).

24. W. G. Golden, D. D. Saperstein, M. W. Severson, and J. Overend, *J. Phys. Chem.* **88**, 574 (1984).
25. A. M. Bradshaw and E. Schweizer, *Advances in Spectroscopy, Spectroscopy of Surfaces* (R. J. H. Clark and R. H. Hester, eds.), Wiley, New York (1988).
26. Z. Xu and J. T. Yates, Jr., *J. Vac. Sci. Technol.* **A8**, 3666 (1990).
27. X. D. Wang, W. T. Tysoe, R. G. Greenler, and K. Truszkowska, *Surf. Sci.* **257**, 335 (1991).
28. S. Chiang, R. G. Tobin, and P. L. Richards, *J. Electron Spectrosc. Relat. Phenom.* **29**, 113 (1983).
29. J. Yarwood, T. Shuttleworth, J. B. Hasted, and T. Nanba, *Nature* **317**, 743 (1984); T. Nanba, *Rev. Sci. Instrum.* **60**, 1689 (1989).
30. C. H. Hirschmugl, G. P. Williams, F. M. Hoffmann, and Y. J. Chabal, *Phys. Rev. Lett.* **65**, 480 (1990).
31. F. M. Hoffmann, K. C. Lin, R. G. Tobin, C. J. Hirschmugl, G. P. Williams, and P. Dumas, *Surf. Sci.* **275**, L675 (1992).
32. D. K. Lambert, *J. Vac. Sci. Technol.* **B3**, 1479 (1985).
33. J. E. Butler, V. M. Bermudez, and J. L. Hylden, *Surf. Sci.* **163**, L708 (1985).
34. J. L. Wragg, H. W. White, and L. F. Sutcu, *Phys. Rev.* **B37**, 2508 (1988).
35. F. M. Hoffmann, *Surf. Sci. Rep.* **3**, 107 (1983).
36. J. E. Reutt-Robey, D. J. Doren, Y. J. Chabal, and S. B. Christman, *J. Chem. Phys.* **93**, 9113 (1990).
37. R. G. Greenler, *J. Vac. Sci. Technol.* **12**, 1410 (1975).
38. J. A. Reffner, *Proc. SPIE-Int. Soc. Opt. Eng.* **1437**, 89 (1991).
39. Z. Jentzich, *Z. Tech. Phys.* **7**, 310 (1926).
40. G. W. Poling, *J. Electrochem. Soc.* **116**, 958 (1969).
41. R. G. Greenler, R. R. Rahn, and J. P. Schwartz, *J. Catal.* **23**, 42 (1971).
42. H. M. Nussenzveig, *Causality and Dispersion Relations*, Academic, New York (1972) p. 43.
43. D. L. Allara, A. Baca, and C. A. Pryde, *Macromolecules* **11**, 1215 (1978).
44a. W. Suëtaka, *J. Spectrosc. Soc. Jpn.* **26**, 251 (1977).
44b. K. Nishikida and R. W. Hannah, *Appl. Spectrosc.* **46**, 999 (1992).
44c. S. Yamazaki, *J. Spectrosc. Soc. Jpn.* **43**, 362 (1994).
45. T. Wadayama and W. Suëtaka, *Surf. Sci.* **218**, L490 (1989).
46. J. P. Hawranek and R. N. Jones, *Spectrochim. Acta* **32A**, 99 (1976).
47. P. N. Schatz, S. Maeda, and K. Kozima, *J. Chem. Phys.* **38**, 2658 (1963).
48. K. Ohta and H. Ishida, *Appl. Spectrosc.* **42**, 952 (1988).
49. K. Ito and W. Suëtaka, unpublished work.
50. W. Suëtaka, *J. Spectrosc. Soc. Jpn.* **16**, 219 (1968).
51. G. Roberts, *Langmuir–Blodgett Films*, Plenum, New York (1990).
52. J. D. Swalen, *Ann. Rev. Mater. Sci.* **21**, 373 (1991).
53. M. C. Petty, in *Polymer Surfaces and Interfaces* (W. J. Feast and H. S. Munro, eds.), Wiley, London (1987), p. 163.
54. J. D. Swalen and J. F. Rabolt, in *Fourier Transform Infrared Spectroscopy*, Vol. 4 (J. R. Ferraro and L. J. Basile, eds.), Academic, New York (1985) p. 283.
55. J. F. Rabolt, F. C. Burns, N. E. Schlotter, and J. D. Swalen, *J. Electron Spectrosc. Relat. Phenom.* **30**, 29 (1983).
56. P-A. Chollet, J. Messier, and C. Rosilio, *J. Chem. Phys.* **64**, 1042 (1976).
57. F. Kimura, J. Umemura, and T. Takenaka, *Langmuir* **2**, 96 (1986).
58. D. L. Allara and R. G. Nuzzo, *Langmuir* **1**, 52 (1985).

59. T. Wadayama, H. Momose, and W. Suëtaka, *J. Spectrosc. Soc. Jpn.* **38**, 192 (1989).
60. R. G. Nuzzo, L. H. Dubois, and D. L. Allara, *J. Amer. Chem. Soc.* **112**, 558 (1990).
61. L. Netzer, R. Isocovici, and J. Sagiv, *Thin Solid Films* **100**, 67 (1983).
62. D. L. Allara and R. G. Nuzzo, *Langmuir* **1**, 45 (1985).
63. R. G. Nuzzo, E. K. Korenic, and L. H. Dubois, *J. Chem. Phys.* **93**, 767 (1990).
64. K. Berrada, P. Dumas, Y. J. Chabal, and P. Dubot, *J. Electron Spectrosc. Relat. Phenom.* **54/55**, 1153 (1990).
65. M. K. Debe, *Appl. Surf. Sci.* **14**, 1 (1982–83).
66. W. Suëtaka, *J. Jpn. Inst. Metals* **32**, 301 (1968).
67. F. P. Mertens, *Surf. Sci.* **71**, 161 (1978).
68. F. P. Mertens, *Corrosion* **34**, 359 (1978).
69. R. Jasinski and A. Iob, *J. Electrochem. Soc.* **135**, 551 (1988).
70. D. K. Ottesen, *J. Electrochem. Soc.* **132**, 2250 (1985).
71. G. W. Poling, *Corrosion Sci.* **10**, 359 (1970).
72. W. Suëtaka and N. Morito, in *Proceedings of the 4th European Symposium on Corrosion Inhibitors,* 67 (1975).
73. S. Thibault, *Corrosion Sci.* **17**, 701 (1977).
74. H. Ishida and R. Johnson, *Corrosion Sci.* **26**, 657 (1986).
75. M. Ohsawa and W. Suëtaka, *Corrosion Sci.* **19**, 709 (1978).
76. G. W. Poling, *J. Electrochem. Soc.* **114**, 1209 (1967).
77. S. Thibault and J. Talbot, *C. R. Acad. Sci.,* Ser. C **272**, 805 (1971).
78. T. Suzuki, T. Sato, and W. Suëtaka, *J. Jpn. Soc. Lubr. Eng.* **24**, 592 (1978).
79. S. A. Francis and A. H. Ellison, *J. Chem. Eng. Data* **6**, 83 (1961).
80. D. L. Wooton and D. W. Hughs, *Lub. Eng.* **43**, 736 (1987).
81. J. T. Yates, Jr., V. S. Smentokowski, and A. L. Linsebigler, *NATO ASI Ser.,* Ser. E **220**, 313 (1992).
82. F. J. Boerio and S. L. Chen, *Appl. Spectrosc.* **33**, 121 (1979).
83. F. J. Boerio and S. L. Chen, *J. Colloid Interface Sci.* **73**, 176 (1980).
84. W. Suëtaka, in *Adhesion Aspects of Polymeric Coatings* (K. L. Mittal, ed.), Plenum, New York (1983) p. 225.
85. J. Mielczarski and J. Leppinen, *Surf. Sci.* **187**, 526 (1987).
86. K. Suzuki, A. Hatta, and W. Suëtaka, *Chem. Lett.* **1973**, 189.
87. H. Saijo, T. Kobayashi, and N. Uyeda, *J. Cryst. Growth* **40**, 118 (1977).
88. M. K. Debe, *J. Appl. Phys.* **55**, 3354 (1984).
89. F. F. So, S. R. Forrest, Y. Q. Shi, and W. H. Steier, *Appl. Phys. Lett.* **56**, 674 (1990).
90. M. K. Debe, R. J. Poirier, and K. K. Kam, *Thin Solid Films* **197**, 335 (1991).
91. C. J. Liu, M. K. Debe, P. C. Leung, and C. V. Francis, *Appl. Phys. Comm.* **11**, 151 (1992).
92. M. K. Debe and K. K. Kam, *Thin Solid Films* **186**, 289 (1990).
93. Y. Toyoshima, K. Arai, A. Matsuda, and K. Tanaka, *Appl. Phys. Lett.* **56**, 1540 (1990); ibid, **57**, 1028 (1990).
94. H. Patel and M. E. Pemble, *J. Cryst. Growth* **116**, 511 (1992).
95. M. J. D. Low and J. C. McManus, *Chem. Comm.* **1967**, 1166.
96. J. T. Yates, Jr. and D. A. King, *Surf. Sci.* **30**, 601 (1972).
97. J. T. Yates, Jr., R. G. Greenler, I. Ratajczykowa, and D. A. King, *Surf. Sci.* **36**, 739 (1973).
98. R. A. Shigeishi and D. A. King, *Surf. Sci.* **58**, 379 (1976).
99. A. M. Bradshaw and J. Pritchard, *Surf. Sci.* **17**, 372 (1969).
100. J. Pritchard and M. L. Sims, *Trans. Faraday Soc.* **66**, 427 (1970).
101. M. L. Kottke, R. G. Greenler, and H. G. Tompkins, *Surf. Sci.* **32**, 231 (1972).

102. M. Ito and W. Suëtaka, *Chem. Lett.* **1973**, 757.
103. E. F. McCoy and R. ST. C. Smart, *Surf. Sci.* **39**, 109 (1973).
104. D. Reinalda and V. Ponec, *Surf. Sci.* **91**, 113 (1979).
105. T. T. Nguyen and N. Sheppard, in *Advances in Infrared and Raman Spectroscopy,* Vol. 5 (R. E. Hester and R. H. J. Clark, eds.), Heyden, London (1978).
106. H-J. Krebs and H. Lüth, *Appl. Phys.* **14**, 337 (1977).
107. J. C. Campuzano and R. G. Greenler, *Surf. Sci.* **83**, 301 (1979).
108. A. M. Bradshaw and F. M. Hoffmann, *Surf. Sci.* **72**, 513 (1978).
109. L. Surnev, Z. Xu, and J. T. Yates, Jr., *Surf. Sci.* **201**, 1 (1988).
110. L. Surnev, Z. Xu, and J. T. Yates, Jr., *Surf. Sci.* **201**, 14 (1988).
111. M. Tüshaus, W. Berndt, H. Conrad, A. M. Bradshaw, and B. Persson, *Appl. Phys.* **A51**, 91 (1990).
112. Y. J. Chabal, S. B. Christman, J. J. Arrecis, J. A. Prybyla, and P. J. Estrup, *J. Electron Spectrosc. Relat. Phenom.* **44**, 17 (1987).
113. J. A. Prybyla, P. J. Estrup, and Y. J. Chabal, *J. Chem. Phys.* **94**, 6274 (1991).
114. Y. J. Chabal, *Phys. Rev. Lett.* **55**, 845 (1985).
115. J. E. Reutt, Y. J. Chabal, and S. B. Christman, *Phys. Rev.* **B38**, 3112 (1988).
116. P. Gardner, R. Martin, M. Tüshaus, and A. M. Bradshaw, *J. Electron Spectrosc. Relat. Phenom.* **54/55**, 619 (1990).
117. P. Gardner, M. Tüshaus, R. Martin, and A. M. Bradshaw, *Surf. Sci.* **240**, 112 (1990).
118. G. Blyholder, *J. Phys. Chem.* **79**, 756 (1975).
119. N. D. Lang, S. Holloway, and J. K. Nørskov, *Surf. Sci.* **150**, 24 (1985).
120a. F. M. Hoffmann and M. D. Weisel, *Surf. Sci.* **253**, L402 (1991); **269/270**, 495 (1992).
120b. M. D. Weisel, J. L. Robbins and F. M. Hoffmann, *J. Phys. Chem.* **97**, 9441 (1993).
121. M. Trenary, K. J. Uram, and J. T. Yates, Jr., *Surf. Sci.* **157**, 512 (1985).
122. J. T. Yates, Jr., M. Trenary, K. J. Uram, H. Metiu, F. Bozso, R. M. Martin, C. Hanrahan, and J. Arias, *Philos. Trans. R. Soc. London* **A318**, 101 (1986).
123. K. J. Uram, L. Ng, M. Folman, and J. T. Yates, Jr., *J. Chem. Phys.* **85**, 2891 (1986).
124. K. J. Uram, L. Ng, and J. T. Yates, Jr., *Surf. Sci.* **177**, 253 (1986).
125. F. M. Hoffmann, N. D. Lang, and J. K. Nørskov, *Surf. Sci.* **226**, L48 (1990).
126. Z. Xu, M. G. Sherman, J. T. Yates, Jr., and P. R. Antoniewicz, *Surf. Sci.* **276**, 249 (1992).
127. M. Trenary, K. J. Uram, F. Bozso, and J. T. Yates, Jr., *Surf. Sci.* **146**, 269 (1984).
128. B. N. J. Persson and R. Ryberg, *Phys. Rev. Lett.* **54**, 2119 (1985).
129. D. H. Ehlers, A. Spitzer, and H. Lüth, *Surf. Sci.* **160**, 57 (1985).
130. Z. Xu, L. Hanley, and J. T. Yates, Jr., *J. Chem. Phys.* **96**, 1621 (1992).
131. Z. Xu, J. T. Yates, Jr., L. C. Wang, and H. J. Kreuzer, *J. Chem. Phys.* **96**, 1628 (1992).
132. R. F. Hicks and A. T. Bell, *J. Catal.* **90**, 205 (1984).
133. T. P. Beebe and J. T. Yates, Jr., *Surf. Sci.* **173**, L606 (1986).
134. B. E. Hayden, K. Kretzschmar, A. M. Bradshaw, and R. G. Greenler, *Surf. Sci.* **149**, 394 (1985).
135. R. G. Greenler, J. A. Dudek, and D. E. Beck, *Surf. Sci.* **145**, L453 (1984).
136. J. Xu, P. Henriksen, and J. T. Yates, Jr., *J. Chem. Phys.* **97**, 5250 (1992).
137. R. G. Greenler, K. D. Burch, K. Kretzschmar, R. Klauser, A. M. Bradshaw, and B. E. Hayden, *Surf. Sci.* **152/153**, 338 (1985).
138. R. G. Greenler, F. M. Leibsle, and R. S. Sorbello, *J. Electron Spectrosc. Relat. Phenom.* **39**, 195 (1986).
139. F. M. Leibsle, R. S. Sorbello, and R. G. Greenler, *Surf. Sci.* **179**, 101 (1987).
140. R. K. Brandt, R. S. Sorbello, and R. G. Greenler, *Surf. Sci.* **271**, 605 (1992).

141. W. F. Banholzer and R. I. Masel, *Surf. Sci.* **137**, 339 (1984).

142. F. M. Hoffmann and J. L. Robbins, *J. Electron Spectrosc. Relat. Phenom.* **45**, 421 (1987).

143. F. M. Hoffmann and M. D. Weisel, *J. Vac. Sci. Technol.* **A11**, 1957 (1993).

144. W. Erley, *J. Electron Spectrosc. Relat. Phenom.* **44**, 65 (1987).

145. R. A. Campbell and D. W. Goodman, *Rev. Sci. Instrum.* **63**, 172 (1992).

146. R. A. Shigeishi and D. A. King, *Surf. Sci.* **75**, L397 (1978).

147. M. W. Lesley and L. D. Schmidt, *Surf. Sci.* **155**, 215 (1985).

148. Th. Fink, J-T. Dath, R. Imbihl, and G. Ertl, *Surf. Sci.* **251/252**, 985 (1991).

149. P. Gardner, R. Martin, M. Tüshaus, and A. M. Bradshaw, *Surf. Sci.* **269/270**, 405 (1992).

150a. F. M. Hoffmann and J. L. Robbins, in: *Proceedings of the 9th International Congress on Catalysis* **3**, 1144 (1988).

150b. C. A. Mims, M. D. Weisel, F. M. Hoffmann, J. H. Sinfelt and J. M. White, *J. Phys. Chem.* **97**, 12656 (1993).

151. L. Ng, K. J. Uram, Z. Xu, P. L. Jones, and J. T. Yates, Jr., *J. Chem. Phys.* **86**, 6523 (1987).

152. J. Xu, X. Zhang, R. Zenobi, J. Yoshinobu, Z. Xu, and J. T. Yates, Jr., *Surf. Sci.* **256**, 288 (1991).

153. J. G. Chen, M. D. Weisel, F. M. Hoffmann, and R. B. Hall, in: *Catalytic Selective Oxidation*, American Chemical Society, Washington, DC (1992).

154. V. A. Burrows, *J. Electron Spectrosc. Relat. Phenom.* **45**, 41 (1987).

155. J. B. Benziger and G. R. Schoofs, *Surf. Sci.* **171**, L401 (1986).

156. F. M. Hoffmann and M. D. Weisel, in: *Surface Science of Catalysis: In-situ Probes and Reaction Kinetics*, ACS Symposium Series, Vol. 482 (D. J. Dwyer and F. M. Hoffmann, eds.), American Chemical Society, Washington, DC (1992).

157. D. H. Chenery, M. A. Chesters, and E. M. McCash, *Surf. Sci.* **198**, 1 (1988).

158. F. M. Hoffmann, *J. Chem. Phys.* **90**, 2816 (1989).

159. C. H. F. Peden, D. W. Goodman, M. D. Weisel, and F. M. Hoffmann, *Surf. Sci.* **253**, 44 (1991).

160. F. M. Hoffmann, M. D. Weisel, and C. H. F. Peden, *Surf. Sci.* **253**, 59 (1991).

161. M. D. Weisel, J. G. Chen, F. M. Hoffmann, Y-K. Sun, and W. H. Weinberg, *J. Chem. Phys.* **81**, 9396 (1992).

162. J. E. Reutt-Robey, D. J. Doren, Y. J. Chabal, and S. B. Christman, *Phys. Rev. Lett.* **61**, 2778 (1988).

163. Y. J. Chabal, *Proc. SPIE* **1145** (Fourier Transform Spectrosc.), 34 (1989).

164. J. E. Reutt-Robey, Y. J. Chabal, D. J. Doren, and S. B. Christman, *J. Vac. Sci. Technol.* **A7**, 2227 (1989).

165. Y. J. Chabal, *Vacuum* **41**, 70 (1990).

166. H. Ibach and D. L. Mills, *Electron Energy Loss Spectroscopy and Surface Vibrations*, Academic, New York (1982).

167. R. G. Tobin and P. L. Richards, *Surf. Sci.* **179**, 387 (1987).

168. D. Hoge, M. Tüshaus, and A. M. Bradshaw, *Surf. Sci.* **207**, L935 (1988).

169. I. J. Malik and M. Trenary, *Surf. Sci.* **214**, L237 (1989).

170. X-D. Wang and R. G. Greenler, *Phys. Rev.* **B43**, 6808 (1991).

171. X-D. Wang, W. T. Tysoe, R. G. Greenler, and K. Truszkowska, *Surf. Sci.* **258**, 335 (1991).

172. X-D. Wang and R. G. Greenler, *Surf. Sci.* **226**, L51 (1990).

173. C. M. Truong, J. A. Rodriguez, and D. W. Goodman, *Surf. Sci.* **271**, L385 (1992).

174. Z. Xu and J. T. Yates, Jr., *Surf. Sci.* **265**, 118 (1992).

175. J. Yoshinobu, T. H. Ballinger, Z. Xu, H. J. Jänsch, M. I. Zaki, and J. T. Yates, Jr., *Surf. Sci.* **255**, 295 (1991).

176. J. A. Rodriguez, W. K. Kuhn, C. M. Truong, and D. W. Goodman, *Surf. Sci.* **271**, 333 (1992).
177. E. Langenbach, A. Spitzer, and H. Lüth, *Surf. Sci.* **147**, 179 (1984).
178. J. Yoshinobu, R. Zenobi, J. Xu, Z. Xu, and J. T. Yates, Jr., *J. Chem. Phys.* **95**, 9393 (1991).
179. X. Guo, J. T. Yates, Jr., V. K. Agrawal, and M. Trenary, *J. Chem. Phys.* **94**, 6256 (1991).
180. A. M. Bradshaw and F. Hoffmann, *Surf. Sci.* **52**, 449 (1975).
181. F. M. Hoffmann and A. M. Bradshaw, Proceedings of the 7th International Vacuum Congress & 3rd International Conference on Solid Surfaces, Vienna (1977) 1167.
182. J. F. Blanke, S. E. Vincent, and J. Overend, *Spectrochim. Acta* **A32**, 163 (1976).
183. W. G. Golden, D. S. Dunn, and J. Overend, *J. Phys. Chem.* **82**, 843 (1978).
184. D. S. Dunn, W. G. Golden, M. K. Severson, and J. Overend, *J. Phys. Chem.* **84**, 336 (1980).
185. T. Wadayama, Thesis, Dr. Eng., Tohoku University (1987).
186. T. Wadayama, K. Monma, and W. Suëtaka, *J. Phys. Chem.* **87**, 3181 (1983).
187. W. G. Golden, D. S. Dunn, and J. Overend, *J. Catal.* **71**, 395 (1981).
188. J. B. Benziger, R. E. Preston, and G. R. Schoofs, *J. Appl. Opt.* **26**, 343 (1987).
189. A. Hatta, H. Matsumoto, and W. Suëtaka, *Chem. Lett.* **1983**, 1077.
190. J. B. Benziger, *Appl. Surf. Sci.* **17**, 309 (1984).
191. A. Hatta, T. Wadayama, and W. Suëtaka, *Anal. Sci.* **1**, 403 (1985).
192. W. G. Golden and D. D. Saperstein, *J. Electron Spectrosc. Relat. Phenom.* **30**, 43 (1983).
193. W. G. Golden, in: *Fourier Transform Infrared Spectroscopy*, Vol. 4 (J. R. Ferraro and L. J. Basile, eds.), Academic, New York (1985) p. 315.
194. B. J. Barner, M. J. Green, E. I. Saez, and R. M. Corn, *Anal. Chem.* **63**, 55 (1991).
195. T. Buffeteau, B. Desbat, and J. W. Turlet, *Appl. Spectrosc.* **45**, 380 (1991).
196. W. G. Golden, D. S. Dunn, C. E. Pavlik, and J. Overend, *J. Chem. Phys.* **70**, 4426 (1979).
197. D. S. Dunn, M. W. Severson, W. G. Golden, and J. Overend, *J. Catal.* **65**, 271 (1980).
198. T. Wadayama, Y. Hanata, and W. Suëtaka, *Surf. Sci.* **158**, 579 (1985).
199. E. M. Stuve, R. J. Madix, and B. A. Sexton, *Surf. Sci.* **119**, 279 (1982).
200. Y. Izumi, in: *Advances in Catalysis,* Vol. 32 (D. D. Eley, H. Pines, and P. B. Weisz, eds.), Academic, New York (1983).
201. A. Hatta, K. Kobayashi, T. Wadayama, and W. Suëtaka, *Chem. Lett.* **1985**, 843.
202. T. Wadayama, T. Saito, and W. Suëtaka, *Appl. Surf. Sci.* **20**, 199 (1984).
203. S. R. Seydmonir, S. Abdo, and R. F. Howe, *J. Phys. Chem.* **86**, 1233 (1982).
204. T. Wadayama, M. Wada, and W. Suëtaka, *Appl. Surf. Sci.* **25**, 231 (1986).
205. M. Kawai, S. Naito, K. Tamaru, and K. Kawai, *Chem. Phys. Lett.* **98**, 377 (1983).
206. W. Knoll, M. R. Philpott, and W. G. Golden, *J. Chem. Phys.* **77**, 219 (1982).
207. K. B. Koller, W. A. Schmidt, and J. E. Butler, *J. Appl. Phys.* **64**, 4704 (1988).
208. T. Wadayama, W. Suëtaka, and A. Sekiguchi, *Jpn. J. Appl. Phys.* **27**, 501 (1988).
209a. T. Wadayama, H. Shibata, T. Ohtani, and A. Hatta, *Appl. Phys. Lett.* **61**, 1060 (1992).
209b. T. Wadayama, H. Shibata, T. Kobayashi, and A. Hatta, *J. Mater. Sci.* **29**, 1041 (1994).
210. T. Wadayama, H. Kayama, A. Hatta, W. Suëtaka, and J. Hanna, *Jpn. J. Appl. Phys.* **29**, 1884 (1990).
211. *Optical Studies of Adsorbed Layers at Interfaces,* Symposium of the Faraday Society, No. 4, London (1971).
212. R. H. Muller, ed., *Advances in Electrochemistry and Electrochemical Engineering,* Vol. 9, *Optical Techniques in Electrochemistry,* Wiley, New York (1973).
213. R. J. Gale, ed., *Spectroelectrochemistry—Theory and Practice,* Plenum, New York (1988).
214. K. Chandrasekaran and J. O'M. Bockris, *Surf. Sci.* **175**, 623 (1986).
215. J. W. Russell, J. Overend, K. Scanlon, M. Severson, and A. Bewick, *J. Phys. Chem.* **86**, 3066 (1982).

216. K. Kunimatsu, W. G. Golden, H. Seki, and M. R. Philpott, *Langmuir* **1**, 245 (1985).
217. F. Kitamura, M. Takahashi, and M. Ito, *Surf. Sci.* **223**, 493 (1989).
218. S. Pons, *J. Electron Spectrosc. Relat. Phenom.* **45**, 303 (1987).
219. D. K. Lambert and R. G. Tobin, *Surf. Sci.* **232**, 149 (1990).
220. N. K. Ray and A. B. Anderson, *J. Phys. Chem.* **86**, 4851 (1982).
221. P. A. Paredes Olivera, E. P. M. Leiva, E. A. Castro, and A. J. Arvia, *J. Electroanal. Chem.* **351**, 65 (1993).
222. S. Holloway and J. K. Nørskov, *J. Electroanal. Chem.* **161**, 193 (1984).
223. D. K. Lambert, *J. Electron Spectrosc. Relat. Phenom.* **30**, 59 (1983).
224. M. R. Philpott, P. S. Bagus, C. J. Nelin, and H. Seki, *J. Electron Spectrosc. Relat. Phenom.* **45**, 169 (1987).
225. K. Kunimatsu, H. Seki, W. G. Golden, J. G. Gordon, II, and M. R. Philpott, *Langmuir* **4**, 337 (1988).
226. D. S. Bethune, A. C. Luntz, J. K. Sass, and D. K. Roe, *Surf. Sci.* **197**, 44 (1988).
227. K. Kunimatsu, H. Seki, W. G. Golden, J. G. Gordon, II, and M. R. Philpott, *Langmuir* **2**, 464 (1986).
228. K. A. B. Lee, K. Kunimatsu, J. G. Gordon, W. G. Golden, and H. Seki, *J. Electrochem. Soc.* **134**, 1676 (1987).
229. K. Chandrasekaran and J. O'M. Bockris, *Surf. Sci.* **185**, 495 (1987).
230. J. Cavilier, R. Faure, G. Guinet, and R. Durand, *J. Electroanal. Chem.* **107**, 205 (1980).
231. D. Aberdam, R. Durand, R. Faure, and F. El-Omar, *Surf. Sci.* **171**, 303 (1986).
232. N. Furuya, S. Motoo, and K. Kunimatsu, *J. Electroanal. Chem.* **239**, 347 (1988).
233. S. Watanabe, J. Inukai, and M. Ito, *Surf. Sci.* **293**, 1 (1993).
234. M. Ito and M. Takahashi, *Surf. Sci.* **158**, 609 (1985).
235. A. Ortega, F. M. Hoffmann, and A. M. Bradshaw, *Surf. Sci.* **119**, 79 (1982).
236. M. W. Severson, W. J. Tornquist, and J. Overend, *J. Phys. Chem.* **88**, 469 (1984).
237. J. Benziger, G. Schoofs, and A. Myers, *Langmuir* **4**, 268 (1988).
238. K. Kretzschmar, J. K. Sass, A. M. Bradshaw, and S. Holloway, *Surf. Sci.* **115**, 183 (1982).
239. G. R. Schoofs and J. B. Benziger, *Surf. Sci.* **192**, 373 (1987).
240. G. R. Schoofs and R. E. Preston, and J. B. Benziger, *Langmuir* **1**, 313 (1985).
241. A. K. Myeres and J. B. Benziger, *Langmuir* **3**, 414 (1987).
242. R. M. A. Azzam and N. M. Bashara, *Ellipsometry and Polarized Light,* North Holland, Amsterdam (1987).
243. For example, M. K. Debe and D. R. Field, *J. Vac. Sci. Technol.* **A9**, 1265 (1991).
244. M. J. Dignam, B. Rao, M. Moscovitz, and R. Stobie, *Can. J. Chem.* **49**, 1115 (1971).
245. J. D. Fedyk and M. J. Dignam, in: *Vibrational Spectroscopies for Adsorbed Species* (A. T. Bell and M. L. Hair, eds.), American Chemical Society, Washington, DC. (1980) p. 75.
246. F. Ferrieu, *Rev. Sci. Instrum.* **60**, 3212 (1989).
247. R. T. Graf, J. L. Koenig, and H. Ishida, *Anal. Chem.* **58**, 64 (1986).
248. J. Bremer, O. Hundri, K. Fanping, T. Skauli, and E. Wold, *Appl. Opt.* **31**, 471 (1992).
249. R. W. Stobie, B. Rao, and M. J. Dignam, *Surf. Sci.* **56**, 334 (1976).
250. J. D. Fedyk, P. Mahaffy, and M. J. Dignam, *Surf. Sci.* **89**, 404 (1979).
251. F. L. Baudais, A. J. Borschke, J. D. Fedyk, and M. J. Dignam, *Surf. Sci.* **100**, 210 (1980).
252. M. J. Dignam and M. D. Baker, *J. Vac. Sci. Technol.* **21**, 80 (1982).
253. H. Ishida, Y. Ishino, H. Buijs, C. Tripp, and M. J. Dignam, *Appl. Spectrosc.* **41**, 1288 (1987).
254. P. R. Mahaffy and M. J. Dignam, *Surf. Sci.* **97**, 377 (1980).
255. P. Mahaffy and M. J. Dignam, *J. Phys. Chem.* **84**, 2683 (1980).
256. N. Blayo, P. Blom, and B. Drévillon, *Physica* **B170**, 566 (1991).

257. N. Blayo and B. Drévillon, *J. Non-Cryst. Solids* **137/138**, 771 (1991).
258. N. Blayo and B. Drévillon, *J. Non-Cryst. Solids* **137/138**, 775 (1991).
259. J. Pritchard and T. Catterick, in: *Experimental Methods in Catalytic Research*, Vol. 3, *Characterization of Surface and Adsorbed Species* (R. B. Anderson and P. T. Dawson, eds.), Academic, New York (1976), p. 281.
260. J. Pritchard, T. Catterick, and R. K. Gupta, *Surf. Sci.* **53**, 1 (1975).
261. K. Horn and J. Pritchard, *Surf. Sci.* **52**, 437 (1975).
262. M. Moskovits, C. J. Hope, and B. Jantzi, *Can. J. Chem.* **53**, 3313 (1975).
263. R. Ryberg, *Surf. Sci.* **114**, 627 (1982).
264. P. Hollins and J. Pritchard, in: *Vibrational Spectroscopies for Adsorbed Species* (A. T. Bell and M. L. Hair, eds.), American Chemical Society, Washington, DC (1980), p. 51.
265. P. Hollins, K. J. Davies, and J. Pritchard, *Surf. Sci.* **138**, 75 (1984).
266. B. E. Hayden and A. M. Bradshaw, *Surf. Sci.* **125**, 787 (1983).
267. B. E. Hayden, K. Prince, D. P. Woodruff, and A. M. Bradshaw, *Surf. Sci.* **133**, 589 (1983).
268. B. E. Hayden, *Surf. Sci.* **131**, 419 (1983).
269. R. Ryberg, *Chem. Phys. Lett.* **83**, 423 (1981).
270. R. Ryberg, *J. Electron Spectrosc. Relat. Phenom.* **54/55**, 65 (1990).
271. J. D. E. McIntyre, *Surf. Sci.* **37**, 658 (1973).
272. A. Bewick and A. M. Tuxford, in *Optical Studies of Adsorbed Layers at Interfaces,* Faraday Society, London (1971), p. 114.
273. W. Paatsch, *Surf. Sci.* **37**, 59 (1973).
274. W. Suëtaka and M. Ohsawa, *Appl. Surf. Sci.* **3**, 118 (1979).
275. R. P. Van Duyne, in: *Chemical and Biochemical Application of Lasers*, Vol. 4 (C. B. Moore, ed.), Academic, New York (1979), p. 101.
276. N. J. Harrick, *Phys. Rev.* **125**, 1165 (1962).
277. G. Samoggia, A. Nucciotti, and G. Chiarotti, *Phys. Rev.* **144**, 749 (1966).
278. A. Hatta, H. Amano, and W. Suëtaka, *Vib. Spectrosc.* **1**, 371 (1991).
279. A. Bewick and S. Pons, in: *Advances in Infrared and Raman Spectroscopy*, Vol. 12 (R. J. H. Clark and R. E. Hester, eds.), Wiley Heyden, Chichester (1985), p. 1.
280. A. Bewick, K. Kunimatsu, B. S. Pons, and J. W. Russell, *J. Electroanal. Chem.* **160**, 47 (1984).
281. A. Bewick, K. Kunimatsu, and B. S. Pons, *Electrochim. Acta* **25**, 465 (1980).
282. B. Beden, *J. Electroanal. Chem.* **345**, 1 (1993).
283. M. I. Lopes, I. Fonesca, P. Olivi, B. Beden, F. Hahn, L. M. Léger, and C. Lamy, *J. Electroanal. Chem.* **346**, 415 (1993).
284. S. Pons, *J. Electroanal. Chem.* **150**, 495 (1983).
285. F. Ozanam and J-N. Chazalviel, *Rev. Sci. Instrum.* **59**, 242 (1988); *J. Electron Spectrosc. Relat. Phenom.* **45**, 323 (1987).
286. J-N. Chazalviel, K. C. Mandel, and F. Ozanam, *Proc. SPIE-Int. Soc. Opt. Eng.* **1575**, 40 (1992).
287. K. Kunimatsu, *Hyomen* **20**, 197 (1982).
288. J-N. Chazalviel and A. Venkateswara Rao, *J. Electrochem. Soc.* **134**, 1138 (1987).
289. D. Laser and M. Ariel, *Anal. Chem.* **45**, 2141 (1973).
290. B. B. Damaskin, O. A. Petrii, and V. V. Batrakov, *Adsorption of Organic Compounds on Electrodes*, Plenum, New York (1971).
291. V. S. Bagotsky, Y. B. Vassiliev, and O. A. Khazova, *J. Electroanal. Chem.* **81**, 229 (1977).
292. A. Weickovski and J. Sobkowski, *J. Electroanal. Chem.* **73**, 317 (1976).
293. B. Beden, C. Lamy, A. Bewick, and K. Kunimatsu, *J. Electroanal. Chem.* **121**, 343 (1981).

294. C. Lamy, F. Hahn, and B. Beden, *Port. Electrochim. Acta* **7**, 435 (1989).
295. E. Morallón, J. L. Vázquez, J. M. Pérez, B. Beden, F. Hahn, J. M. Léger, and C. Lamy, *J. Electroanal. Chem.* **344**, 289 (1993).
296. T. Solomun, *Surf. Sci.* **176**, 593 (1986).
297. B. Beden, A. Bewick, K. Kunimatsu, and C. Lamy, *J. Electroanal. Chem.* **142**, 345 (1982).
298. K. Kunimatsu, *J. Phys. Chem.* **88**, 2195 (1984).
299. B. Beden, N. Collas, C. Lamy, J. M. Léger, and V. Solis, *Surf. Sci.* **162**, 789 (1985).
300. B. Beden, C. Lamy, N. R. de Tacconi, and A. J. Arvia, *Electrochim. Acta* **35**, 691 (1990).
301. R. O. Lezna, K. Kunimatsu, T. Ohtsuka, and N. Sato, *J. Electrochem. Soc.* **134**, 3090 (1987).
302. A. V. Rao, F. Ozanam, and J-N. Chazalviel, *J. Electrochem. Soc.* **138**, 153 (1991).
303. K. C. Mandel, F. Ozanam, and J-N. Chazalviel, *J. Electroanal. Chem.*, in press.
304. M. J. Weaver, D. S. Corrigan, P. Gao, D. Gosztola, and L-W. H. Leung, *J. Electron Spectrosc. Relat. Phenom.* **45**, 291 (1987).
305. D. S. Corrigan and M. J. Weaver, *Langmuir* **4**, 599 (1988).
306. K. Ashley, M. Lazaga, M. G. Samant, H. Seki, and M. R. Philpott, *Surf. Sci.* **219**, L590 (1989).
307. K. Ashley, F. Weinert, M. G. Samant, H. Seki, and M. R. Philpott, *J. Phys. Chem.* **95**, 7409 (1991).
308. D. S. Corrigan, J. K. Foley, P. Gao, S. Pons, and M. J. Weaver, *Langmuir* **1**, 616 (1985).
309. S. Pons, S. B. Khoo, A. Bewick, M. Datta, J. J. Smith, A. S. Hinman, and G. Zachmann, *J. Phys. Chem.* **88**, 3575 (1984).
310. T. Wadayama, Y. Momota, A. Hatta, and W. Suëtaka, *J. Electroanal. Chem.* **289**, 29 (1990).
311. J. O'M. Bockris, J. L. Carbajal, B. R. Scharifker, and K. Chandrasekaran, *J. Electrochem. Soc.* **134**, 1957 (1987).
312. B. R. Scharifker, M. A. Habib, J. L. Carbajal, and J. O'M. Bockris, *Surf. Sci.* **173**, 97 (1986).
313. P. P. Karat and N. V. Madhusudana, *Mol. Cryst. Liq. Cryst.* **36**, 51 (1976).

3

Internal Reflection Spectroscopy

1. Attenuated Total Reflection Spectroscopy

1.1. Basic Principles and Instrumentation

When a beam of light is incident at an angle of incidence, ϕ_1, on a plane interface from a transparent medium 1, part of the light is reflected at the interface and travels back in the plane of incidence, as shown in Fig. 1. In this specular or regular reflection, the angle of reflection equals the angle of incidence. If medium 1 is optically denser than medium 2, i.e., $n_1, > n_2$, this reflection is called internal reflection. The remaining part of the incident light is refracted into medium 2 (supposed to be transparent) according to Snell's law

$$n_1 \sin \phi_1 = n_2 \sin \phi_2 \tag{1}$$

where ϕ_2 is the angle of refraction and is larger than ϕ_1 because $n_1 > n_2$. When the sine of an incident angle is equal to $n_{2\,1}$ ($= n_2/n_1$), the incident angle is called the critical angle and is denoted by ϕ_c. Beyond the critical angle, the incident light does not travel in medium 2 and is totally reflected. In the total reflection, the reflected light, whose amplitude is equal to that of the incident beam, combines with the incident light to yield a standing wave, as shown in Fig. 2. The oscillating electric field of the standing wave penetrates into medium 2. The penetrating field is evanescent and does not propagate in the rarer medium, because its amplitude decays exponentially with distance from the interface.

If the rarer medium 2 has absorption bands, the intensity of the evanescent field and consequently that of the reflecting light is attenuated at the wavelength of the absorption of medium 2. It follows that the reflected light bears information about the absorption spectrum of medium 2. The observation of the IR beam totally reflected at the interface between a transparent and optically dense

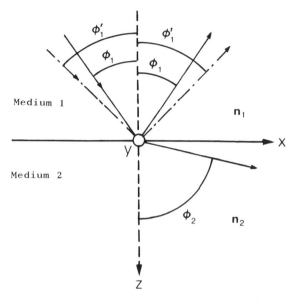

Figure 1. Reflection and refraction of light at the interface between an optically denser medium 1 and a rarer medium 2. The light beam shown with a dot-and-dash line is totally reflecting: $\phi_1 < \phi_c$, $\phi_1' > \phi_c$ and ϕ_c is the critical angle.

material called an internal reflection element (IRE) and a sample brought into close contact with the IRE yields an IR absorption spectrum of the sample.

The observation of one type of absorption spectrum in the internal reflection arrangement was developed by Harrick[1] and Fahrenfort[2] and is called attenuated total reflection (ATR) spectroscopy[2] or internal reflection spectroscopy.[3] Attentuated total reflection spectroscopy is a feasible method for the observation of surface layers and very strongly absorbing materials whose absorption spectra are difficult to measure using the conventional transmission method. Since the ATR method is a well established technique, and excellent texts and reviews have been published,[3–8] selected examples of applications of ATR spectroscopy and a newly developed technique (in which an enhanced surface field is utilized for the observation of metal/sample interfaces at a high sensitivity) are the focus of this chapter. Attenuated total reflection measurements combined with the electrode potential modulation technique are included in Section 3 of Chapter 2.

Since the amplitude of the evanescent oscillating electric field decays exponentially with distance from the interface, the ATR method provides an

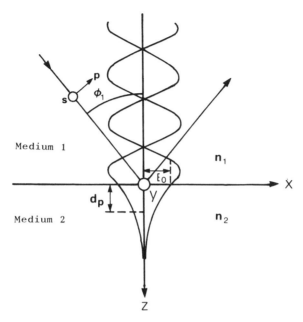

Figure 2. Schematic drawing of the standing wave established in medium 1 upon total reflection of light and the evanescent wave in medium 2.

absorption spectrum of the surface layer of the sample, where the evanescent wave has substantial strength. The depth of penetration d_p, defined by Harrick as the distance where the amplitude of the evanescent wave falls to e^{-1} of its value at the interface (E_0 in Fig 2), is given by[3]

$$d_p = \lambda_1/2 \; \pi(\sin^2 \phi_1 - n_{21}^2)^{1/2} \qquad (2)$$

where λ_1 ($= \lambda_0/n_1$) is the wavelength of the incident light in medium 1 and λ_0 is its wavelength in a vacuum. The relative penetration depth, d_p/λ_1, at the germanium/air interface is plotted versus angle of incidence in Fig. 3. A theoretical calculation has shown that a surface layer of d_p in thickness contributes 86% and a layer having a thickness $d_p/2$ from the interface contributes 63% of the measured total intensity.[9] Figure 3 shows that, if germanium is used as the IRE, the penetration depth in air is about λ_1 or less except for near-critical angle incidence. Equation (2) indicates that the depth of penetration increases with decreasing refractive index of the IRE, and increases also in proportion to the wavelength of the incident light. Roughly speaking, the

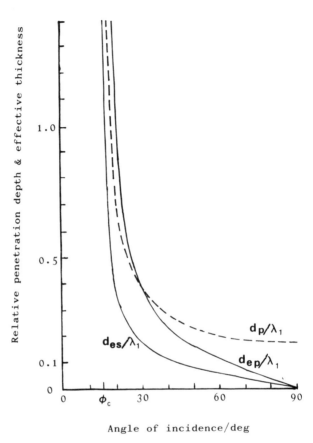

Figure 3. Calculated relative penetration depth and effective thickness of the evanescent wave for a germanium/air interface. d_{es} and d_{ep} are the effective thicknesses for s- and p-polarized light, respectively.

spectra of surface layers, whose thickness is less than some μm, are obtained with the IR ATR method, provided the sample is in perfect contact with the reflecting plane of the IRE. Materials having high refractive indices and that are transparent in the spectral region of interest are used for the IRE. Typical materials used for the IRE and their properties are collected in Table 1. Several semiconductors are conveniently used as internal reflection elements, but a rise in temperature brings about an increase in absorption due to free carriers, i.e.,

Table 1. Typical Materials for IRE in the IR Region

Material	Usable wavelength	Refractive index	Remarks
Al_2O_3	0.2–4.5 μm	1.8	Very hard, brittle, chemically stable
As_2SE_3	1.0–12	2.8	Brittle, poisonous, soluble in alkaline solution
ZnSe	0.5–15	2.4	Hard, brittle, soluble in acid
AgCl	0.4–20	2.0	Very soft, easily scratched, sensitive to UV light
KRS-5	0.5–35	2.4	Soft, slightly soluble in water, poisonous, may be inhomogenous
ZnTe	0.6–52(?)	2.7	Decomposes in acidic solution
Si	1.1–7, 20–far IR	3.4	Hard, brittle, cheap, chemically stable
Ge	2.0–11.5	4.0	Hard, brittle, chemically stable

the use of germanium as the IRE at temperatures higher than 330 K may give rise to serious problems.

The sensitivity of ATR spectroscopy is generally lower in the observation of species adsorbed on the surface of the IRE than that of oblique incidence reflection spectroscopy in the observation of adsorbates on highly reflective metal surfaces. However, multiple reflections are easily applied in the ATR observation, because the intensity of IR light does not decrease upon reflection like it does in the oblique incidence reflection method. Harrick has shown that the signal strength increases in proportion to the number of reflections, as will be described below. The ATR method is also utilized in the visible and near-ultraviolet region. Since the penetration depth is one order of magnitude smaller in these spectral regions than in the infrared region, the behavior of species present in the closest vicinity of the surface of the IRE that is working, for instance, as an electrode can be observed *in situ*. This subject has been dealt with in Refs. 5 and 6.

The ATR method provides us with a type of absorption spectrum that is not identical with the ordinary absorption spectrum measured with the conventional transmission method. When the sample has a thickness much larger than the penetration depth, the relative intensity of absorption bands increases with wavelength. The refractive index of the sample shows the anomalous dispersion in the region of absorption, resulting in distortion of the spectrum, as will be described below. Furthermore, the electric field of the evanescent wave changes in intensity depending upon the polarization state of the incident IR beam. The absorption intensity, consequently, changes with the polarization state of the incident light in the ATR measurement, even if the sample is isotropic. It follows that the absorbance measured using unpolarized light cannot be used for the quantitative analysis without appropriate correction.[9] In the qualitative study, the use of perfectly polarized light, preferably p-polarized light for reasons of high S/N ratio, is appropriate.

The penetration depth changes remarkably with the angle of incidence, as shown in Fig. 3. When an IR beam is incident at an angle near the critical angle, the denominator of Eq. (2) becomes very small. This seems to show that a fairly thick layer of the sample could be observed by light incident at an angle near the critical angle. However, the refractive index of the sample shows a sharp change due to the anomalous dispersion in the region of absorption, as shown in Fig. 4. This figure also shows that the strong absorption band accompanies a considerable anomalous dispersion.

Let us suppose that the IR beam is incident at an angle somewhat larger than the critical angle for the nonabsorbing region of the sample. Then the IR beam may be totally reflected in the wavenumber region where the refractive index of the sample is smaller than the broken line in Fig 4. However, the penetration depth changes remarkably in the region of absorption because of the change in n_{21}, and in the hatched region where the penetration depth is infinite, the IR beam propagates into the sample. The spectra measured at this incident angle shows severe distortion.

If the incident angle used is slightly higher than the critical angle for the maximum refractive index at the anomalous dispersion, a sizable change of penetration depth still results, and the recorded absorption band shows a distortion in band shape and a shift of band peak. When the sample has strong absorptions, therefore, the incident angle must be considerably larger than the critical angle for the nonabsorbing region of the sample in order to obtain a spectrum without significant distortion. Further, it should be noted that, if the incident IR beam is convergent, half of the beam is incident at angles smaller than the incident angle of the beam center. For these reasons, the IR beam is generally incident at angles somewhat larger than the critical angle for the absorbing region of the sample, and consequently the mean penetration depth is usually less than λ_1.

The change in the incident angle results in the change in the strength of the electric field at the interface between the IRE and the sample, in addition to the change in the penetration depth. The area sampled by the IR beam increases in inverse proportion to cosine of the incident angle. The observed intensity of the absorption bands of the sample changes depending upon all of these factors. Harrick introduced a concept of effective thickness for comparing the spectra obtained with the ATR method to ordinary absorption spectra taken by the transmission method.[3]

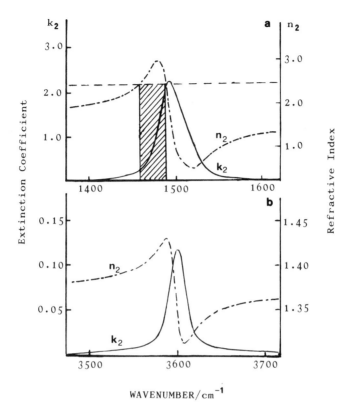

Figure 4. Optical constants generated using damped harmonic oscillator models. (a) Very strongly absorbing band, (b) moderately absorbing band.

In the transmission measurement, the intensity, I_d, of IR light after passing through a very thin film of thickness d is expressed, neglecting the reflection at the surfaces, as

$$I_d = I_0 \exp(-\alpha d) \tag{3}$$

where I_0 and α are the intensity of the incident IR light and the absorption coefficient of the film, respectively. Since $1 \gg \alpha d$, except for very strongly absorbing bands, Eq. (3) can be reduced in a good approximation to

$$I_d/I_0 = 1 - \alpha d \tag{4}$$

In the ATR measurement of a single reflection, the reflectance R can be written as

$$R \equiv I/I_0 = 1 - a \tag{5}$$

where a is the reflection loss per reflection due to the absorption of the evanescent wave. Analogous to Eq. (4), Eq. (5) can be rewritten as

$$R = 1 - \alpha d_e \tag{6}$$

where $d_e \ (= a/\alpha)$ is defined as an effective thickness. For multiple reflections, the reflectance $R(N)$ is given by

$$R(N) = (1 - \alpha d_e)^N \tag{7}$$

where N is the number of reflections. If $1 \gg \alpha d_e$, this equation can be reduced to

$$R(N) = 1 - N \alpha d_e \tag{8}$$

This equation shows that the reflection loss increases in proportion to the number of reflections.

When the sample has a thickness much larger than the penetration depth and an IR beam of unit strength is propagating in medium 1 toward the interface, the effective thickness is related to the penetration depth by Eq. (9), as long as medium 2 is weakly absorbing,[3]

$$d_e = n_{21} E_0^2 d_p / 2 \cos \phi_1 \tag{9}$$

where E_0^2 is the intensity of the oscillating standing wave at the interface and in medium 2. The factor $1/\cos \phi_1$ corresponds to the change in area sampled by the IR beam.

In the coordinate system shown in Fig. 2, where the z-axis is perpendicular to the interface and the y-axis is perpendicular to the plane of incidence and points outward from the figure plane, the x and z components of the oscillating electric field are derived from p-polarized light, and the y component from s-polarized light. When polarized light of unit intensity is incident on the interface, the intensity of these components of the electric field at $z = 0$ and in medium 2 are[1,3]

$$E_{y0}^2 = 4 \cos^2 \phi_1 / (1 - n_{21}^2) \tag{10}$$

$$E_{x0}^2 = \frac{4(\sin^2 \phi_1 - n_{21}^2) \cos^2 \phi_1}{(1 - n_{21}^2)\{(1 + n_{21}^2)\sin^2 \phi_1 - n_{21}^2\}} \tag{11}$$

$$E_{x0}^2 = \frac{4 \sin^2 \phi_1 \cos^2 \phi_1}{(1 - n_{21}^2)\{(1 + n_{21}^2)\sin^2 \phi_1 - n_{21}^2\}} \tag{12}$$

In medium 1, the intensities of the x and y components are identical with those in medium 2. However, E_{z0}^2 reduces by a factor of n_{21}^4 in medium 1. Fig. 5 illustrates the calculated intensity of the electric field at a germanium/air interface and in the air phase. The calculated electric field is stronger than the incident beam except at incident angles larger than 60° and becomes largest at the critical angle. The largest value is 4 irrespective of IRE material for s-polarized light and is $64(= 4 \times n_{12}^2)$ for p-polarized light. Fig. 5 also shows that p-polarized light gives only an electric field oscillating normal to the interface at critical angle incidence. This is also the case in the visible region,

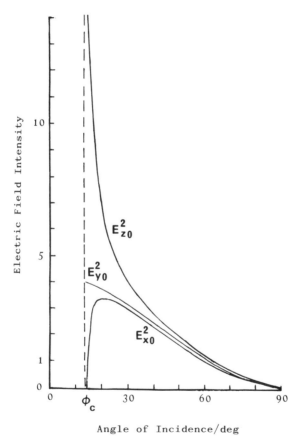

Figure 5. Calculated electric field intensity at a germanium/air interface, but in the air generated during the incidence of IR light of unit strength: E_{x0} and E_{z0} come from p-polarized light and E_{y0} from s-polarized light.

and the exciting laser beam must be incident at the critical angle for obtaining maximum sensitivity in total reflection Raman observation. For the same reason, the s-polarized laser beam is often introduced at this angle for generating the collective electron resonance on an island film of silver evaporated on the reflecting plane of the IRE.

Taking into account $E_{0p}^2 = E_{x0}^2 + E_{z0}^2$ and $E_{0s}^2 = E_{y0}^2$, and substituting E_0^2 and d_p in Eq. (9) by Eqs. (10)-(12) and (2), one has the effective thickness of isotropic samples for s- and p-polarized light, respectively:

$$d_{es} = \frac{n_{21}\cos\phi_1\,\lambda_1}{\pi\,(1 - n_{21}^2)(\sin^2\phi_1 - n_{21}^2)^{1/2}} \tag{13}$$

$$d_{ep} = \frac{n_{21}\cos\phi_1(2\sin^2\phi_1 - n_{21}^2)\lambda_1}{\pi\,(1 - n_{21}^2)\{(1 + n_{21}^2)\sin^2\phi_1 - n_{21}^2\}(\sin^2\phi_1 - n_{21}^2)^{1/2}} \tag{14}$$

The relative effective thicknesses, d_e/λ_1, at a germanium/air interface are plotted versus angle of incidence in Fig 3.

Equations (6), (13), and (14) show that the reflection loss arising from the absorption of the sample increases in proportion to the effective thickness, which increases, in turn, in proportion to wavelength. The relative intensity of absorption bands observed by the ATR method increases, therefore, with wavelength.

Attenuated total reflection IR spectroscopy is often employed in the investigation of thin films on various semiconductors. In this measurement, medium 2 is generally very thin and the evanescent field has a sizable intensity even on the outside of medium 2, as shown in Fig. 6. Since the thickness d of the surface film is very small in comparison with the penetration depth, the electric field intensity can be considered constant throughout the film. The effective thickness is, then, given by[3,10]

$$d_e = n_{21}\,E^2 d/\cos\phi_1 \tag{15}$$

The intensities of the electric field, E^2, in the film for both the light polarizations are given by[10]

$$E_s^2 = 4\cos^2\phi_1/(1 - n_{31}^2) \tag{16}$$

$$E_p^2 = \frac{4\cos^2\phi_1\{(1 + n_{32}^2)\sin^2\phi_1 - n_{31}^2\}}{(1 - n_{31}^2)\{(1 + n_{31}^2)\sin^2\phi_1 - n_{31}^2\}} \tag{17}$$

Inserting Eqs. (16) and (17) into Eq. (15), one obtains the effective thicknesses for s- and p-polarized light:

$$d_{es} = 4n_{21}d\cos\phi_1/(1 - n_{31}^2) \tag{18}$$

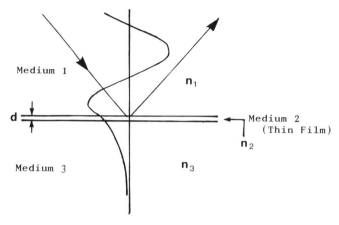

Figure 6. Schematic drawing of the standing and evanescent waves in a three-phase system: medium 2 is a thin film.

$$d_{ep} = \frac{4n_{21}\cos\phi_1\{(1 + n_{32}^4)\sin^2\phi_1 - n_{31}^2\}d}{(1 - n_{31}^2)\{(1 + n_{31}^2)\sin^2\phi_1 - n_{31}^2\}} \tag{19}$$

These equations have been proven to be good approximations, provided $2\pi d/\lambda_1 < 0.1$ and $k_2/n_2 < 0.1$.[10] One may notice that the film thickness, d, appears in Eqs. (18) and (19) in place of λ_1 in Eqs. (13) and (14). It turns out that the intensities of the absorption bands of the film increase with film thickness and are independent of the wavelength of the incident light, in contrast to bulk samples. In addition, Eqs. (18) and (19) imply that the critical angle is related to n_{31} instead of n_{21} in the case of bulk samples. The spectra of thin films measured with the ATR method are, therefore, not seriously distorted even when the IR beam is incident at an angle near the critical angle. In consequence, the ATR spectra of thin films approximate ordinary absorption spectra,[3] though they are not free from the band distortion due to the anomalous dispersion of the refractive index of the film. It should be added further that materials optically rarer than the sample can be employed as the IRE in the ATR measurement and the resultant decrease in refractive index may give rise to an increase in the effective thickness.[3] Equations (16)–(19) show that the intensity of the electric field in the film, and consequently the effective thickness, increase generally, but not always, when the refractive index of medium 3 approaches that of the IRE, although the light beam must be incident at a large angle and the number of reflections decreased.[6] A numerical calculation shows that when, in the ATR observation of a SiO_2 thin film on a Si plate, the Si substrate is used as the IRE, a thick coating of AgBr on the SiO_2 film causes increases in the relative effective thickness (d_e/d), from 1.7 to 2.1 for s-polarized light of near critical angle incidence and from 5 to 25 for p-polarized light. Olsen and Shimura noticed a remarkable enhancement of electric field in a film of SiO_2 on Si in the ATR configuration where Ge was used as the reflection plate.[11].

Various types of internal reflection elements are employed in ATR measurement.[3,6] A few examples of the internal reflection elements are shown in Fig 7. The hemicylindrical element in Fig. 7a is generally used in fundamental investigations, because the angle of incidence can be changed continuously over a wide range. However, observations requiring high sensitivity are difficult with this single reflection element. The ATR measurements are gen-

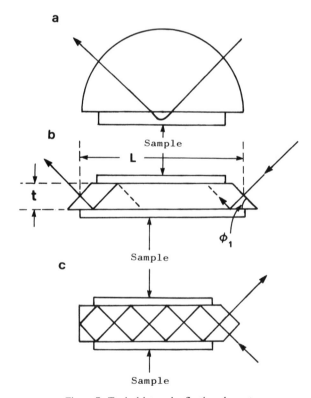

Figure 7. Typical internal reflection elements.

erally performed with multiple reflection plates. The simplest types of the multiple reflection plates are shown in Fig. 7b and c. The number of reflections (strictly speaking, the number of contact reflections[8]), N, of single-pass reflection elements such as that in Fig. 7b contacting the sample on both sides is calculated from the equation

$$N = L \cot \phi_1 / t \qquad (20)$$

where L and t are the length and thickness of the IRE, respectively. The double-pass plate in Fig. 7c is often employed for the observation of electrode/electrolyte interfaces in the vertical position. However, internal reflection plates of this type are not adequate for ensuring a sufficient energy throughput using the circular beam of FTIR spectrometers. Recently, a cylindrical internal reflection cell was developed as an attachment to FTIR spectrometers for the investigation of reactions in solution.[12] Its internal reflection element is a rod-shaped crystal with cone-shaped ends. When an element made of ZnSe is used, the measurement can be performed above 700 cm^{-1}.

Increasing the number of reflections is an efficient way to enhance the signals from the sample. Increasing the number of reflections, however, inevitably demands the enlargement of the sample and IRE plate. Thin reflection plates are then used for avoiding an excessive increase in size of the sample. The use of thin reflection plates also enables an extension of the observable region to wavenumbers where the reflection plate is weakly absorbing.[13]

Sample preparation is easier in the ATR method than in the conventional transmission method. The dissolution of the sample in solvents or the reduction of thickness of the solid samples for obtaining a spectrum exhibiting absorption bands of adequate intensity is not necessary in the ATR observation. In the ATR observation, however, care must be taken to bring the sample into perfect contact with the reflecting plane of the IRE. Good contact is easily obtained for liquid samples and elastic solid materials having plane surfaces.

In the measurement of hard solid samples, the use of a sample having a plane surface is important for obtaining good contact. However, samples having perfectly plane surfaces are not always available, and stress is generally applied to an elastic sheet placed on the back of the sample so as to obtain good contact between the sample and the IRE. However, soft internal reflection plates may be flawed and hard and brittle plates may cleave under the applied stress. Sharp projections and edges of the sample, therefore, must be abraded before measurement. Another technique commonly used to achieve good contact is the insertion of a liquid or soft solid substance optically denser than the sample and transparent in the wavenumber region of interest between the sample and the reflecting plate. A thin plate of silver chloride, for example, may be placed between the sample and the IRE, and an adequate stress is put on the sample system to ensure good contact. In practice, the AgCl plate must be out of contact with metallic materials, because AgCl corrodes ordinary metals except silver.

Generally speaking, perfect contact between a hard solid sample and the IRE is difficult to attain. In the investigation of solid surfaces, excellent results may be obtained by the use of the sample itself as the internal reflection element, provided the solid material is optically dense and transparent over a wide wavenumber region.[8] For instance, a shallowly oxidized silicon plate was utilized in the investigation of silicon oxide surfaces. The analysis of compounds formed on the corroded surfaces of semiconductors has been carried out using the semiconductors as the IREs.[8] Investigation of adsorption on semiconductor surfaces has been carried out in the same way.

The absorption intensity and accordingly the signal-to-noise ratio of the spectrum obtained increases in proportion to the effective thickness in the ATR measurement. Since the effective thickness increases in proportion to n_{21}, as can be seen from Eq. (9), the use of optically less dense materials as reflection elements results in an increase of the S/N ratio. At the same time, the decrease in refractive index of the IRE gives an increase in the penetration depth, according to Eq. (2). The measurement of the spectrum of thin surface layers at a high S/N ratio can be best done by multiple reflection measurement using an internal reflection element having a high refractive index such as Ge. When thin surface layers of polymers are investigated, the spectrum of the surface layer is generally obtained by the subtraction of the spectrum of the substrate polymer from the sample spectrum, as will be described below.

The penetration depth also changes depending upon the incident angle. The incident angle can be changed continuously with the reflection element shown in Fig. 7a, and in practice, the penetration depth can be changed continuously in the range of $0.2\lambda_1$–λ_1. The change in composition and molecular orientation of the sample with distance from the surface could be observed over a wide range, in principle, by the use of reflection elements of various refractive indices at diverse incident angles. However, the observed signals are generally very weak in the single reflection measurement at high incident angles. On the other hand, the change in incident angle is generally restricted to a narrow range if multiple reflection elements are used. Consequently, continuous observation of the depth dependence of the spectrum is not very easy in practice.

The incidence of p-polarized light induces an electric field in the sample oscillating in the x and z directions, and for s-polarized light the field oscillates in the y direction, as mentioned above. As can be seen from Fig. 5, the electric field intensities, E_{x0}^2, E_{y0}^2 and E_{z0}^2, are comparable with one other except near the critical angle incidence. This is in contrast to specular reflection high resolution electron energy loss spectroscopy and oblique incidence IR reflection spectroscopy of metal surfaces, which are sensitive only to the dynamic dipole moment normal to the surface. This is a big advantage of ATR–IR spectroscopy in the investigation of molecular orientation. The investigation of anisotropy in thin films and surface layers can be done using linearly polarized or partially polarized IR radiation.[14] The molecular orientation in uni-axially oriented samples can be determined by alternate incidence of *p*- and *s*-polarized IR beams. For the investigation of bi-axially oriented samples, a rotatable internal reflection attachment was developed.[15] In the ATR measurement, a longitudinal electric field associated with the standing wave is present at the interface, in sharp contrast to the transmission measurement, and absorption bands, which are lacking in the transmission spectra, may appear in the ATR spectra.[16]

1.2. Applications

1.2.1. Semiconductor Surfaces

The majority of investigations of semiconductor surfaces have been done using the semiconductor itself as the IRE. Adsorption on semiconductor surfaces can be observed with a high sensitivity in this arrangement, because multiple reflections of 100 or more can be attained without excessive difficulty using thin semiconductor plates. Harrick observed the formation of C–H and O–H bonds on a mechanically polished Si surface.[3] Becker and Gobeli first observed a broad absorption band of the Si–H stretching vibration on a Si surface exposed to atomic hydrogen. They also monitored the formation of Si–H bonds arising from the hydrogen ion bombardment of the surface of a Si plate.[17]

The adsorption of hydrogen on well-defined Si surfaces is of considerable interest not only from the scientific standpoint, but also in relevance to the technological applications, and the vibrational spectra of hydrogen adsorbed on Si were observed with HREELS.[18,19] Chabal studied the adsorption of hydrogen on Si[100] and Ge[100] surfaces by ATR–IR spectroscopy, showing that, whereas mono- and dihydride species are formed on the Si surface, only GeH species are present on the Ge surface.[20] Recently, however, Lu and Crowell observed the presence of GeH_2 GeH_3 species on germanium surfaces.[21] Although hydrogen-covered semiconductor surfaces have generally been prepared by the adsorption of atomic hydrogen on clean semiconductor surfaces in ultrahigh vacuum, the hydrogen-covered surface can also be prepared by chemical removal of the surface oxide in solution. The removal of the chemically formed oxide on Si and Ge surfaces in aqueous HF solutions produces very inert surfaces showing a very low recombination velocity.[22] Thus prepared, hydrogen-covered Si surfaces are contaminant-free and stable at room temperature and hence have become a subject of various investigations.[23]

Chabal *et al.* investigated HF-treated Si[111] surfaces with polarized ATR–IR spectroscopy. The observation of Si–H stretching vibrations has shown that the chemical bonds are terminated at the surface with mono-, di-, and trihydride groups. The sharpness of the absorption bands and the complexity of the obtained spectra imply that the surface is well ordered but is microscopically rough.[24] The IR measurement has shown also that mono-, coupled mono-, di-, and trihydride species are present on Si[100] and stepped Si[111] surfaces as well as the flat Si[111] surface. The presence of the variety of surface species indicates the microscopic roughness of these surfaces. The results also suggest the probable interaction of the trihydride with the steps on the vicinal Si[111] surface, which gives rise to the splitting of the degenerate asymmetric Si–H stretching vibration and the enhancement of the dynamic dipole moment of the vibration perpendicular to the steps.[25] Hydrogen-terminated Si surfaces can also be prepared by HF treatment under the irradiation of ultraviolet light. The presence of SiH and SiH_2 groups on Si[100] has been detected by observation of ATR–IR spectra.[26]

The above-mentioned microscopically rough Si[111] surface can be drastically modified by changing the pH of the HF etching solution. When basic solutions (pH = 9–10) are used, the etched Si[111] surface shows a very sharp Si–H stretching band of monohydride oriented normal to the surface, as shown in Fig. 8. The spectrum of the Si[111] surface etched in a dilute HF solution shows broad bands of monohydride (M), dihydride (D), and trihydride (T) depicted by a broken line in this figure. This means the Si surface etched in basic HF solution is ideally terminated with silicon monohydride.[27]

The formation of the atomically flat monohydride-terminated Si[111] surface may be attributed to the preferential etching, in which defect structures are removed. Vicinal Si[111] surfaces (stepped Si[111] 9° surfaces) were used for examining the preferential etching by buffered HF solutions.[28] Since the polarized spectra of the surfaces etched in basic HF solutions showed very sharp absorption bands, the terrace and step species could easily be identified and thus the etching process could be examined. The data suggest that isolated adatoms on [111] planes are removed at the highest rate and that the edge atoms, which constitute kinks, are dissolved faster than those at the steps, resulting in atomically straight steps. These works were reviewed by Chabal.[23]

The Si/SiO_2 interface formed by thermal oxidation of Si was found to be

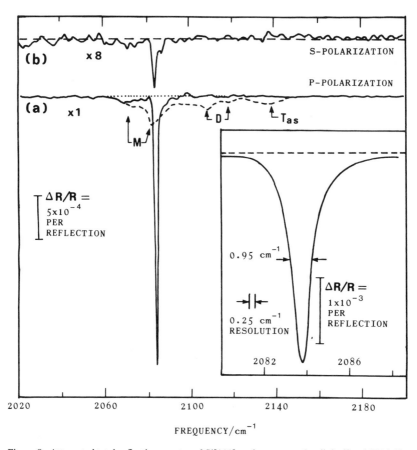

Figure 8. Attenuated total reflection spectra of Si[111] surfaces trerated wtih buffered HF (pH = 9–10) (solid lilne) and with dilute HF (dashed line). Inset: High-resolution spectrum of Si[111] surface treated with a basic HF solution (pH = 9–10). Incident angle = 45° [Reprinted from G. S. Higashi, Y. J. Chabal, G. W. Trucks, and K. Raghavachari, *Appl. Phys. Lett.* **56**, 656 (1990)].

much smoother than that formed by chemical oxidation.[29,20] Attenuated total reflection IR spectroscopy was then utilized to investigate the dependence of the structure of the hydrogen-covered Si surface on the nature of the initial surface oxide. The HF etching of the thermally oxidized Si surfaces produced much flatter surfaces than chemically oxidized surfaces, and the IR spectra exhibited a strikingly sharp Si–H stretching band having a halfwidth of 0.066 cm^{-1} at 40 K, when the oxide formed thermally on Si[111] surfaces was removed by buffered HF (pH = 5) followed by etching in a 40% ammonium fluoride solution.[31] The sharp band indicates the formation of a remarkably homogeneous H/Si[111] surface.

The morphology of hydrogen-terminated Si[111] and Si[100] surfaces prepared by etching in HF solutions was investigated with HREELS and ATR–IR spectroscopy. The EELS observation has shown no signal relating to possible impurities such as F, OH, C, O, or hydrocarbons, confirming that the surfaces are completely terminated with hydrogen. Infrared and EELS observations show that, whereas Si[111] surfaces etched in basic HF solutions are atomically flat, the Si[100] surfaces remain rough independent of pH, although the surface morphology varies depending upon pH.[32]

The very sharp Si–H stretching vibration band appearing in the IR spectra of ideally H-terminated Si[111] surfaces is also a subject of fundamental study. The temperature dependence of intensity, width, and peak frequency of the sharp Si–H band was investigated over the range 130–560 K. The data can be explained by a weak anharmonic coupling to a silicon surface phonon, and a strong anharmonic coupling is also presumed in the Si–H bending vibration at 637 cm^{-1}.[33] The inhomogeneous broadening of the absorption bands of surface species comes from the diversity of intermolecular distances between adsorbates, surface impurities, and defects such as steps, edges, dislocation ends, and kinks.[34] Only small inhomogeneous broadening effects may be expected for the Si–H band of the ideally H-terminated Si[111] surfaces. In fact, the Si–H shows exceptionally small broadening and an asymmetric band shape with a tail on the lower frequency side at low temperatures. On the other hand, the homogeneous broadening is calculated to be very small at low temperatures.[33] The observed broadening thus should be attributed mostly to the inhomogeneous one. The origin of the asymmetric inhomogeneous broadening has been studied theoretically, and two major sources of the broadening, both of which are related to the dynamic dipole coupling within finite size domains, have been identified.[35]

Amorphus films of a–Si:H contain molecular hydrogen that is responsible for some of the properties of the amorphous silicon. Infrared absorption of the molecular hydrogen trapped in the a-Si:H film, which stems from the dynamic dipole moment induced upon collisions with the surrounding molecules and atoms and is very weak, could be observed by the multiple reflection ATR method.[36] The molecular hydrogen was further studied with nuclear magnetic resonance and calorimetry, in addition to the ATR–IR technique. The results show that the molecular hydrogen is trapped in microbubbles of about 1 nm in diameter under high pressure at room temperature.[37]

The reaction of GaAs with NO$_2$ and NO was investigated in ultrahigh vacuum with relevance to a technique to produce dielectric layers on semiconductor surfaces. NO is adsorbed on GaAs[110] both dissociatively and molecularly at 90 K. The reaction products, N_2O and N_2O_2 were found at this temperature and arsenic nitride was found above 200 K.[38] The oxidation of GaAs by NO_2 has been studied extensively with the ATR–IR technique. The absorption bands of oxides formed on GaAs change in relative intensity and bandwidth depending upon the procedure employed for the oxidation. Lenezycki and Burrows have attributed the observed changes to the difference in the short-range structure of the oxide. They have extended the measurement to *in situ* observation of anodic oxidation of GaAs and found that, when the electrochemically formed oxide becomes thick, its structure does not differ so much from those formed in gaseous oxidation.[39]

The study, in which molecular beam techniques and transport measurements were used in parallel with IR spectroscopy, has demonstrated that NO_2 is adsorbed dissociatively on GaAs[110] to show Ga–O(782 cm^{-1}) and As–O (990 cm^{-1}) bands even at 77 K. The weak absorption appearing at 715 cm^{-1} has been ascribed to oxygen atoms migrating into the bulk of the GaAs crystal and causing the evolution of the bulk conductance. The increased exposure to NO_2 at 100 K resulted in the appearance of physisorbed N_2O_4. On the other hand, when GaAs was exposed to oxygen at 300 K, only a weak band appeared at 765 cm^{-1} (782 cm^{-1} when observed at 100 K), suggesting a very low sticking probability for molecular oxygen. In the presence of a hot filament, oxygen molecules were dissociated and atomic oxygen was adsorbed on GaAs, exhibiting a broad band at 850 cm^{-1} attributable to heterogeneously adsorbed oxygen.[40]

The ATR–IR method is quite feasible for the determination of the composition and the detection of impurities in various films on semiconductor surfaces formed by etching, corrosion, oxidation, and vapor deposition. Annapragada *et al.* successfully applied this method to the analysis of the carbon content in GaAs thin films grown on GaAs substrates using a thin Ge reflection plate.[13] The etching of a semiconductor surface leaves various reaction products on the surface. In an early use of ATR–IR spectroscopy, Beckmann applied this method to the analysis of the films formed on Ge during etching in a HF/HNO$_3$ solution. He found hydride, hydroxide, and vitreous oxides, in addition to weak bands of the etchant.[41] Several compounds are often found on the surfaces of Si, Ge, and GaAs etched in hydrofluoric acid buffered with ammonium fluoride as a consequence of insufficient rinsing. The presence of physisorbed ammonium salts was detected on all of the semiconductors investigated. The physisorbed salts slowly corrode the semiconductor surfaces.[42,43] Impurities in oxide films formed on Si were investigated, and the absorption bands of hydride and hydroxide were detected.[44]

1.2.2. Thin Films and Polymer and Electrode Surfaces

Polarized ATR–IR spectroscopy is a convenient technique for the investigation of molecular orientation in thin films and in surface layers.[14] However, in some papers, early, inexact formulae were used for the analysis of alignment. The interpretation of the polarized ATR spectra of well-oriented molecular films was made using correct equations by Haller and Rice[45] and Maoz and Sagiv.[46] Takenaka *et al.* studied the orientation of stearic acid molecules in Langmuir–Blodgett films formed on Ge internal reflection elements.[47,48] The polarized ATR–IR method has been also applied to the investigation of the alignment of polymer molecules in the surface layer of stretched polypropylene,[14,49] nylon[50] and polystyrene.[51]

Dielectric films evaporated on optical components may adsorb water vapor in the atmosphere and deteriorate in quality as a coating material. Thin films of MgF_2 and cryolite were evaporated on KRS-5 internal reflection plates and the adsorbed quantity of water was estimated from the spectra recorded after exposure to water vapor. Thin MgF_2 films of 200 nm or less in thickness adsorbed a much larger quantity of water than did cryolite films of the same thickness, and a highly porous structure was found for thin MgF_2 films.[52]

The cylindrical internal reflection cell was utilized to observe *in situ* the events occurring at the α-FeOOH/aqueous solution interface.[53] The ATR spectra of α-FeOOH thin films are in fairly good agreement with absorption spectra obtained with the conventional transmission method, showing that the oxyhdroxide layers are very thin. Both positively and negatively charged α-FeOOH species induce an ordered structure in adjacent water layers. Phosphate ion perturbs the structure of water in the interphase to a much higher degree than "inert" anions such as NO_3^-, Cl^-, and ClO_4^-. Electropolymerization is used to produce polymer coatings on conductive substrates. The oxidative coupling reaction of 2,6-dimethylphenol on iron electrodes in alkaline methanol solutions was observed *in situ* by the use of a multiple reflection element. The anodic shift of electrode potential first induces the removal of adsorbed methanol, and the deposition of polymer produced in the electrochemical oxidation of phenol begins after the partial removal of adsorbed methanol and increases rapidly. Thus the deposited polymer coating is stable and not removed by a cathodic shift of potential. When 2,4-dimethylphenol was used instead of the 2,6-derivative, the polymerization did not take place and methanol was oxidized, leaving formate ions adsorbed on the electrode.[54]

The depth profile analysis of surface layers thinner than about 2 μm can be done nondestructively by changing both the material of the IRE and the incident angle. The relative concentration of carbonyl and hydroperoxy groups formed in polypropylene films under the irradiation of UV light was determined using this procedure. The concentration was estimated from the measurement of the intensities of the bands of the groups in question relative to those of the C–H stretching vibration band at various penetration depths.[55] The irradiation with UV light also induces the change of structure in the polymer surface layer. The change induced in polypropylene films was investigated and an increase in crystallinity was found in the surface layer.[56] The concentration of oxygen as a function of the distance from the surface was determined for polyethylene films treated with chromic acid.[57]

The composition of surface layers of polymer materials are conveniently determined by the ATR–IR method. The concentration of poly(dimethylsiloxane) in the surface layer of Cardiothane 51, a biomaterial, was determined from the analysis of difference spectra between Cardiothane and its main component, polyurethane.[58] The identification of functional groups present on carbon fiber surfaces is of significance for fundamental understanding of the fiber/matrix interaction. Attenuated total reflectance spectroscopy has been applied to the characterization of rayon-based graphitized carbon fiber surfaces. The data suggest the presence of carboxylic acid, ester, lactone, enol, and quinone moieties on the oxidized surfaces. The change resulting from various functionalization reactions of fibers has been also investigated.[59] Attenuated total reflectance IR spectroscopy has been employed also for the determination of optical constants[60] and of the thickness of surface layers.[61] Optical constants of an isotropic material are also determined from a single ATR spectrum by the use of the Kramers–Kronig transformation.[62-64]

2. Absorption Enhancement by the Presence of a Metal Layer

2.1. Metal Over- and Underlayer ATR Spectroscopy

2.1.1. Metal Overlayer (Otto type) Configuration

The combination of an IRE and a metal plate (or film) is used in the Otto and Kretschmann ATR configurations (Fig. 9) for the excitation of the surface electromagnetic wave of a surface plasmon. However, the enhancement of absorption bands also takes place without exciting a surface electromagnetic wave. This section deals with the absorption enhancement arising from mechanisms other than the excitation of surface electromagnetic waves, and the enhancement taking place in the Otto type configuration is treated first.

The absorption enhancement due to the increase in refractive index of the incident medium is expected from the approximate Eq. (9) of Chapter 2:

$$(\Delta R/R_0)_p = -4n_1^3 n_2 \, \alpha d \sin^2 \phi_1 / (n_2^2 + k_2^2)^2 \cos \phi_1$$

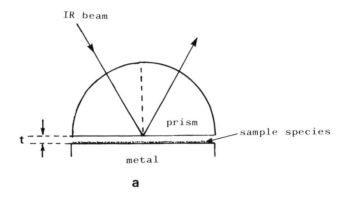

a

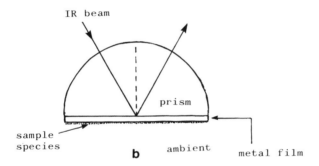

b

Figure 9. Schematic representation of the (a) Otto and (b) Kretschmann configurations.

where n_1 is the refractive index of the incident medium. In an early work on oblique incidence reflection measurement, the enhancement of absorption bands in the Otto type configuration was observed using a KBr prism. Figure 10 shows that the absorption bands of a polymer coating on a tin-free steel plate increase in intensity upon changing the incident medium from air to KBr ($n = 1.53$). A very small amount of Nujol was applied to the film surface to obtain good contact with the prism. Enhancement factors ranging from 3 to 4 were obtained for various thin films on Ni, Al, and steel surfaces.[65] Although the enhancement factors are in agreement with the value calculated from the approximation equation, the equation becomes invalid when the refractive index of the incident medium increases further.

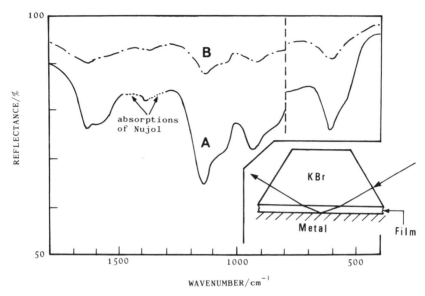

Figure 10. Infrared reflection spectra of a surface film on a tin-free steel measured (A) with and (B) without a KBr prism; p-polarized light was incident. Inset is the arrangement of the sample employed in the measurement (adapted from Ref. 65).

Ishida *et al.* have treated theoretically the absorption enhancement caused by the evaporation of a silver layer over the sample film on the IRE.[66,67] Figure 11 shows the calculated signal strength, ΔR (= $R - R_0$), as a function of refractive index of the incident medium.[66] This figure indicates that the use of Ge ($n = 4$) as the incident medium (or the internal reflection element) brings about an enhancement of more than 5 times that of the absorption bands of the sample films in the Otto type configuration. This simulation has also shown that the absorption enhancement takes place upon incidence of p-polarized light and that the absorption intensity continues to increase with increase in thickness of the silver layer until the layer reaches a thickness of about 20 nm.

The enhancement of absorption bands has been confirmed experimentally[66] in the Otto type configuration using the combination of a poly(vinyl acetate) (PVAc) or cadmium arachidate L-B film and a Ge prism. Figure 12 shows that the increase in the refractive index from 1 to 4 brings about an enhancement of absorption bands of about 5 times, in accordance with the theoretical prediction.[66] Badilescu *et al.* deposited silver overlayers on sample

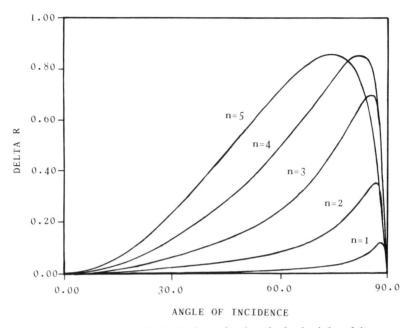

Figure 11. Signal strength (ΔR) calculated as a function of refractive index of the transparent incident medium. The ΔR is due to the carbonyl stretching vibration band of a 5-nm poly(vinyl acetate) film under the Ag overlayer [Reprinted from Y. Ishino and H. Ishida, *Appl. Spectrosc.* **42**, 1296 (1988)].

films in their ATR–IR investigation of water adsorption by tungsten oxide (WO_3) and obtained enhanced signals from the sample species.[68]

This metal-overlayer ATR technique is appropriate for the observation of very thin films present on the surfaces of semiconductor crystals, provided the film remains unchanged during the deposition of the metal film. On the other hand, the application of a prism of optically dense material onto the thin film on a metal substrate is effective for the enhancement of weak absorption bands of the film, as long as the prism can be brought into good contact with the sample film.

Graphite has high optical constants and can be compared with metal. Sellitti *et al.* investigated the effect of overcoating with a graphite layer on the ATR spectrum of an underlying polymer film. The coating with graphite yields an enhancement of signals from the polymer as a result of the increase in the electric field strength oscillating normal to the reflecting plane of the IRE, although it is accompanied by the distortion of absorption bands, the extent of

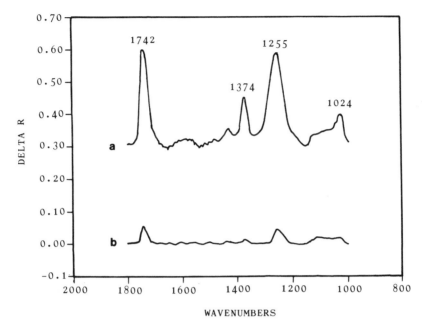

Figure 12. (a) Infrared ATR spectrum of a PVAc film of 10 nm in thickness under a 20-nm Ag overlayer. Spectrum (b) is an external reflection spectrum of a PVAc film of the same thickness formed on Ag taken at an incident angle of 75°; *p*-polarized light was incident in both of the measurements. Prism: Ge [Reprinted from Y. Ishino and H. Ishida, *Appl. Spectrosc.* **42**, 1296 (1988)].

which varies depending upon the distance between the IRE and the graphite layer.[69]

Hartstein *et al.*,[70] Kamata *et al.*,[71] and Badilescu *et al.*[72] observed that the infrared absorptions from organic molecules in thin films formed on IREs are enhanced by the deposition of a Ag or Au overlayer. Suzuki *et al.* showed the intensity enhancement of SCN⁻ ions adsorbed on Au particles in the system ZnSe-prism/LiF-layer (KSCN-bearing)/Au-island-film.[73] The intensity enhancement in these systems is accounted for by the collective electron resonance that is treated in the next section. Sigarev and Yakovlev observed enhanced IR absorption spectra of mixed L-B films of stearic acid and barium stearate using the metal-overlayer multiple reflection ATR technique.[74] The absorption intensity of the L-B films increased with the thickness of the evaporated Ag layer, in accordance with the results obtained by Ishida *et al.*[66] However, absorption spectra could be obtained upon incidence of *s*-polarized light as well, and the intensity of the reflected beam decreased notably when the thickness of the Ag layer increased beyond 10 nm. This result cannot be explained as being due to the effect of the increase in refractive index of the incident medium.

2.1.2. Metal Underlayer (Kretschmann) Configuration

The Kretschmann ATR configuration (Fig. 9) is used for the observation of surface-enhanced Raman spectra of species adsorbed on coinage metal (Ag, Au, and Cu) films. The enhancement of the signal from species on the metal surfaces is also obtained in the infrared region, though the enhancement is modest in comparison with that of surface-enhanced Raman scattering (SERS). The mechanism of the enhancement in the infrared region is not fully understood yet, but the enhancement is utilized for the detection of trace species on solid surfaces.

Hartstein *et al.* first found the enhancement of IR absorption bands of monomolecular layers, which stems from the presence of a thin Ag or Au over- or underlayer, in their ATR observations. They estimated an enhancement factor of 20 and attributed the enhancement to the transverse collective electron resonance of the metal island film.[70] Hatta *et al.*[75] observed the ATR–IR spectra of p-nitrobenzoic acid (PNBA) on Ag underlayers and found that the enhancement was restricted to vibrations belonging to the A_1 species of p-nitrobenzoate ion formed in the reaction of PNBA with silver.

Figure 13 shows polarized ATR spectra of a PNBA thin film of 12 nm thickness formed on an Ag film 20 nm thick.[76] No band was located when s-polarized light was incident, as Fig. 13a shows, but in contrast, the incidence of p-polarized light gave the spectrum in Fig. 13b, in which two distinct bands appear at 1350 and 1390 cm^{-1}, assignable to the NO_2 and COO^- symmetric stretching vibrations, respectively. Figure 13c was recorded with the oblique incidence reflection measurement of the same sample. A striking difference is seen between Figs. 13b and c. The external reflection spectrum is in good agreement with the spectrum of PNBA measured with the conventional transmission method, showing that the sample layer is composed mainly of PNBA molecules. The ATR spectrum, by contrast, is attributable to p-nitrobenzoate ion, which is oriented with its C_2 axis normal to the metal surface, provided the charge transfer described below does not occur. No band of PNBA molecule appears in the ATR spectrum, notwithstanding that PNBA is the chief component of the sample. This seems to mean that only the species directly bound with the metal surface give enhanced absorption in the Kretschmann configuration and that the absorption enhancement takes place by a short-range mechanism.

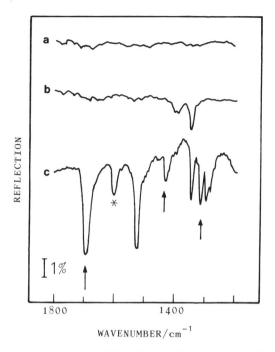

Figure 13. Polarized spectra of a 12-nm PNBA film on a 20-nm Ag layer. ATR spectra (a, b) were taken with either (a) *s*- or (b) *p*-polarized radiation, both incident at an angle of 70°. External reflection spectrum (c) was measured with *p*-polarized light incident at an angle of 80°. * indicates the band due to phenyl ring and ↑ the bands of carboxyl groups [Reprinted from Y. Suzuki, M. Osawa, A. Hatta, and W. Suëtaka, *Appl. Surf. Sci.* **33/34**, 875 (1988)].

Figure 13 shows that the absorption bands of *p*-nitrobenzoate ion adsorbed on Ag are enhanced by a short-range mechanism. However, there is evidence indicating the existence of a fairly long-range enhancement mechanism. Figure 14 shows the intensity increase in the carbonyl stretching band of poly(ethyl cyanoacrylate) with film thickness.[77] Figure 14a shows the intensity of the films formed on a Ag underlayer, and Fig. 14b without the Ag underlayer. The intensity increases steeply in the presence of the Ag layer, until the polymer layer reaches a thickness of about 5 nm, and continues to grow thereafter at the same rate as in the regular ATR measurement. The nonlinear intensity increase can be explained by assuming a mechanism, which stems from the Ag layer and is short-range in comparison to the normal ATR spectrum, superimposed over the normal ATR mechanism, in which the intensity increases in proportion to

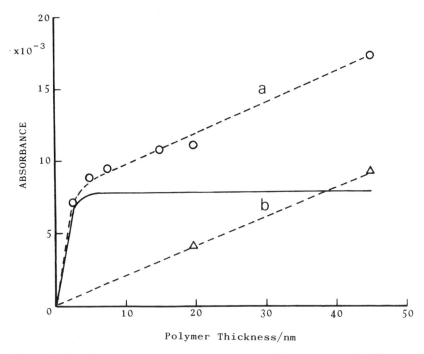

Figure 14. Absorption intensity of the carbonyl stretching band of poly(cyanoacrylate) films as a function of film thickness: (O) observed intensity with a 5-nm Ag underlayer; (Δ) without Ag layer. The solid line is the difference between **a** and **b,** and corresponds to the absorption enhancement due to the Ag layer [Reprinted from A. Hatta, N. Suzuki, Y. Suzuki, and W. Suëtaka, *Appl. Surf. Sci.* **37**, 299 (1989)].

the film thickness because the thickness is much smaller than the penetration depth.

Kamata *et al.* found in their ATR measurement of Langmuir–Blodgett films of stearate that the enhancement from the Ag layer is salient for the stearate ions directly adsorbed on the Ag surface but extends to the third molecular layer.[71] These data imply that the enhanced field extends to a distance of ~5 nm or a little more from the metal surface. In addition, when *m*- or *o*-nitrobenzoic acid is used as the sample molecules instead of the *p*-derivative, the absorption of free molecules not directly attached to the Ag surface is enhanced by the presence of the Ag layer.[78,79] These observations show that the observed absorption enhancement comes from a fairly long-range mechanism that is electromagnetic in nature. Hence the existence of two different intensity enhancement mechanisms is plausible in the absorption enhancement by the presence of coinage metal films.

The enhancement changes markedly in changing the thickness of the metal underlayer. When the Ag layer has a thickness of around 5 nm and consists of fine particles, the incidence of either of p- and s-polarized light brings about the enhancement by a factor of 300, while an enhancement of about 60 times is obtained solely in the incidence of a p-polarized IR beam if the thickness of the Ag layer increases to ~20 nm.[78]

The electric field strength E_v^2 ($v = x, y, z$) generated within Ag films of various thicknesses on the Ge prism base was calculated using a three-layer model.[76] Because the films used in the above-mentioned measurements had island structure or were very rough even if they were continuous, a homogeneous and parallel sided layer of Ag–air composite was assumed in the calculation instead of a homogeneous Ag layer. The dielectric constant of the composite layer was estimated from Bruggeman's effective-medium approximation.[80] Figure 15 shows the calculated effective field strength generated upon incidence of an IR beam of unit cross section. The coordinate system shown in Fig. 1 is used in the calculation. The factor $1/\cos\theta$ comes from the surface area sampled by the incident beam. The observed absorption intensity of the NO_2 symmetric stretching vibration of p-nitrobenzoate ion and the band at 1005 cm^{-1} of Ag(I)-mercaptobenzothiazole complex formed on a continuous Ag film is also shown as a function of the angle of incidence in this figure.[81] The ATR spectra of the mercaptobenzothiazole complex were taken with p-polarized light. Qualitative agreement is seen between the measured absorption intensity and the calculated field strength, showing that the fields E_x and E_y in the composite layer are responsible for the absorption enhancement. At the same time, the agreement indicates that the electric field oscillating parallel to the prism base brings about the absorption enhancement. This means the enhanced absorption should be observed in the transmission measurement as well, which has been confirmed by separate observations.[71,81,82]

The charge-transfer interaction of adsorbates with the metal surface may give rise to the short-range enhancement. Devlin and Consani have predicted the possibility of the charge-transfer enhancement of IR bands of species adsorbed on metal surfaces.[83] When the vibronically produced charge-transfer takes place between the adsorbate and the metal, the absorption intensity of the totally symmetric vibrations of the adsorbate is remarkably enhanced by the electron density oscillation of the charge-transfer, even if the atomic displacement in the vibrational modes is parallel to the metal surface. It should be noted that the charge transfer may also take place between adsorbed species.

Wadayama *et al.* observed the behavior of species generated electrochem-

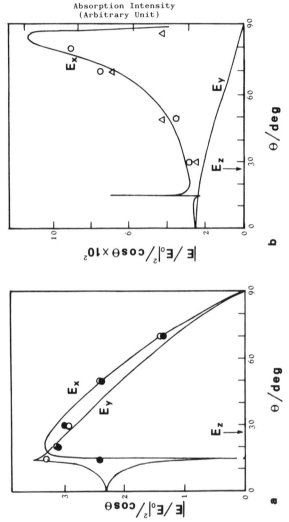

Figure 15. Calculated effective field strength as a function of incident angle (θ) and the observed intensity of the band at 1350 cm^{-1} of p-nitrobenzoate (●, ○) and the 1005 cm^{-1} band of Ag(I)-mercaptobenzothiazole (△). The x -and z-components of the field are parallel and perpendicular to the prism base, respectively, and come from p-polarized light, and the y-component comes from s-polarized radiation. (a) 5 nm thick PNBA film on a 5.5-nm Ag layer (● and ○ measured with p- and s-polarized light, respectively), (b) 12-nm PNBA film on a 20-nm Ag layer and Ag(I)-mercaptobenzothiazole complex film at a Ag/aqueous solution interface. The intensity of the bands was very weak when s-polarized light was incident, and the intensity observed with s-polarized light is not shown in (b). E_0 amplitude of the incident radiation [Reprinted from Y. Suzuki, M. Osawa, A. Hatta, and W. Suētaka. *Appl. Surf. Sci.* **33/34**, 875 (1988) and A. Hatta, Y. Suzuki, T. Wadayama, and W. Suētaka, *Appl. Surf. Sci.* **48/49**, 222 (1991)].

ically from tetracyanoethylene (TCNE), using a Au electrode in the Kretsch-mann configuration, and located absorption bands assignable to vibrations of infrared-inactive A_g species of TCNE^{2-} ion and TCNE^{-} anion radical adsorbed on the electrode.[84] Pons *et al.* also located an IR band assignable to the totally symmetric vibration of TCNE^{-} anion radical absorbed on a Pt electrode.[85]

The totally symmetric C–N stretching vibration band of SCN^{-} ions ad-sorbed on a Au electrode in the Kretschmann configuration increased in in-tensity at highly negative potentials, where the reflectivity of the electrode increased appreciably, showing that the excitation of collective electron reso-nance is unlikely.[86] The number of adsorbed SCN^{-} ions should be smaller at the negative potentials than at more positive electrode potentials, and the intensity of the C–N stretching band of SCN^{-} ions adsorbed on a gold electrode continued to increase in a positive scan of the electrode potential in an external reflection IR measurement.[87] The charge-transfer enhancement is presumably working on the vibration of the SCN^{-} ion adsorbed at negative potentials on the Au electrode. Furthermore, Osawa and Ikeda found that all the bands that were enhanced in the IR spectra of *p*-nitrobenzoate ion adsorbed on Ag island films belong to the totally symmetric A$_1$ species of the ion with C$_{2v}$ symmetry.[88] Consequently, the charge-transfer mechanism plays probably an important role in the short-range enhancement.

The ATR spectrum in Fig. 13b shows absorption bands of only A$_1$ species of benzoate ion. The bands are probably enhanced by the short-range mecha-nism of the charge-transfer, which is induced, in turn, by the electric field in the composite layer. However, a fairly long-range enhancement mechanism is probably working on the rough metal surface, as mentioned above. Then the absence of the signals from PNBA molecules, the main component of the film, in the ATR spectrum requires some elucidation. It may be explained by assum-ing the formation of a multilayered benzoate ion film. On the other hand, the band due to the COO^{-} group of the benzoate ion is hard to locate in the external reflection spectrum (Fig. 13c). The reflection spectrum is induced by the electric field outside the composite layer, which is orthogonal to the field inside the composite layer. If only a few benzoate ions are present outside the compo-site layer, the signal of the ions is faint in the reflection spectrum.

Using a model of a metal particle coated with a uniform layer of a damped harmonic oscillator, Osawa *et al.* showed that the short-range enhancement can be attributed to the IR absorption of the metal particles induced by the vibration of adsorbed species.[89,90] If this mechanism is working, the absorption intensity increases with the thickness of the coating molecular layer until the layer reaches a thickness of 2 ~ 3 nm.[89] Furthermore, this classical electromagnetic model does not account for the absorption enhancement of infrared-inactive vibrational modes. Hence, the mechanism proposed by Osawa *et al.* will take part in the fairly long-range enhancement.

The plasma resonance by metal particles lying on solid surfaces brings about strong local electromagnetic fields in the visible and near-infrared region, giving rise to surface-enhanced spectroscopy such as SERS and the enhancement of photochemical reactions, as will be described in Chapter 6. The plasma resonance may occur in the infrared region and bring about a fairly long-range enhancement, if the metal spheroids in the film have extremely large aspect ratios and are oriented with their long axes parallel to the substrate surface.[91] This was not the case in the above-mentioned measurements. In reality, however, the films consist of particles of various sizes and shapes, and the averaged aspect ratio increases with the film thickness.[92,93] Furthermore, the interaction between metal particles may give rise to the shift of the resonance frequency to a long wavelength. Osawa and Ikeda have inferred that when metal particles are strongly aggregated, a cooperative resonance (collective electron resonance) may be excited in the IR region by the electric field parallel to the substrate surface.[88] Continuous but rough metal films can be modeled with metal spheroids on a thin smooth substrate. The excitation of the localized electromagnetic field may, therefore, be expected on the rough metal films.

Nakao and Yamada investigated the absorption enhancement of polymers brought into contact with various metal (Ag, Ni, Pd, and Pt) films evaporated on a KRS-5 hemicylindrical prism.[94] The strongest absorption signal was obtained with the use of the Ni film, and the enhancement was most remarkable when the metal film was continuous and smooth. They attributed the absorption enhancement to the frustrated total reflection cavity[3] as a result of the multiple reflections of the IR beam in the metal film.

The presence of a coinage metal film thus gives rise to the enhancement of IR absorption bands of species present at a distance shorter than ~5 nm from the metal surface. As a consequence, if a metal film is used as the electrode in the Kretschmann configuration, the signals from the solvent and solute species are remarkably reduced in intensity in the resulting spectra. The behavior of species on the electrode, therefore, can be observed *in situ* without bothersome interference from the solvent and the solute species.[84,86,95–97] Either of the external reflection methods combined with various modulation techniques or SNIFTIRS is mostly used for the infrared investigation of metal electrode surfaces. The thin layer electrochemical cell must be used in these external reflection measurements so as to reduce the enormous absorption of the solvent. However, the thin layer cell gives rise to serious drawbacks, as described in Sections 2.1.3.2 and 3.1 of Chapter 2. In contrast to the external reflection method, bulk solution cells can be used in the measurement with the Kretschmann configuration, and the measurement can be carried out free from draw-backs such as slow diffusion of reaction products and reactant species, a difference between the applied potential and the actual one, and the uneven distribution of current density on the electrode. An example of the electrolytic cells employed in the ATR measurement is shown in Fig. 16.

The polarization modulation method (Section 2.1 of Chapter 2) can be applied to the observation of metal electrode surfaces in the Kretschmann configuration for improving the S/N ratio, because only the p-polarized IR beam gives rise to the absorption enhancement when the metal film is con-tinuous, as mentioned above.

The behavior of CN^- ions on a Ag electrode was observed *in situ* with the polarization modulation method.[96] The C–N stretching frequency changes with the change in electrode potential, as shown in Fig. 17. The potential dependence of the absorption peak of CN^- adsorbed on a smooth Ag disk electrode was investigated by Kunimatsu *et al.*,[98,99] using the polarization modulation external reflection method. The data obtained in their work are also shown in Fig. 17. The results obtained by these two different methods are in good agreement with one other. However, strong bands appeared in the external reflection measurement at 2080 and 2136 cm^{-1}, assignable to cyanide ion and $Ag(CN)_2^-$ complex ion, respectively. These bands have been ascribed to solute species, because the peak frequency of the bands remained constant indepen-dent of the electrode potential. The former band was missing in the spectrum measured with the Kretschmann configuration. A weak feature could be located at 2143 cm^{-1} at a potential −0.6 V (versus Ag/AgCl) in the ATR measurement. This band can be assigned to the $Ag(CN)_2^-$ ion, but its frequency is higher by

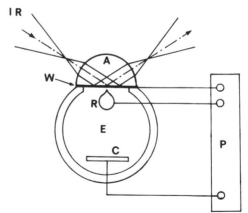

Figure 16. Schematic drawing of the electrolytic cell employed in the *in situ* observation of species on the me-tal electrode in the Kretschmann configuration. A = ATR prism, W = working electrode, R = reference electrode, C = counter electrode, E = electrolyte solution, P = potentiostat [Reprinted from A. Hatta, Y. Chiba, and W. Suëtaka, *Appl. Surf. Sci.* **25**, 327 (1986)].

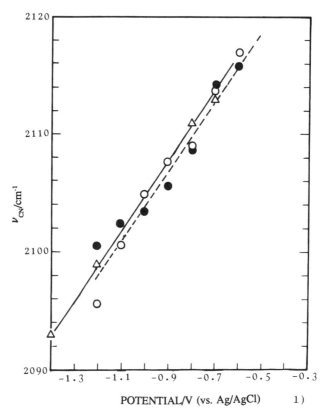

Figure 17. Electrode potential dependence of the CN stretching frequency of CN$^-$ ions adsorbed on a Ag electrode: data from solutions containing 10^{-4} M NaCN (O), 2.5×10^{-4} M KAg (CN)$_2$ (●), and external reflection data by Kunimatsu et al.[98] (△) [Reprinted from A. Hatta, Y. Sasaki, and W. Suëtaka, J. Electoanal. Chem. **215**, 93 (1986)].

8 cm^{-1} than that of the solute species.[100] This difference may indicate that the weak feature observed in the Kretschmann configuration does not come from the complex ion in solution but from the ion adsorbed on the electrode.

The observation of the Au electrode surface in contact with an aqueous solution containing SCN$^-$ ion has shown the formation of a probable precursor of soluble thiocyanato complex of gold on the electrode.[860] Pursuing these works, Matsuda et al. have observed in situ the electrochemical reduction of spontaneously organized films (see Section 1.4 of Chapter 2) of p-nitrothiophenol formed on a Ag electrode.[101].

The enhanced IR absorption spectra can also be obtained with the external reflection configuration when a thin metal film is evaporated onto the sample layer on semiconductor or dielectric surfaces. Nishikawa *et al.* demonstrated the sensitivity increase in the reflection spectra obtained by the introduction of a metal over- or underlayer. The IR absorption spectra of thin organic layers of subnanometer thickness could be measured with fairly good contrast by the reflection method.[102] The measurement of surface-enhanced IR absorption spectra was also made for the detection of lactate in human perspiration in relevance to fatigue after hard physical labor. Absorption bands of triglycerides, fatty acids, and their esters were detected in addition to the features assignable to lactate in the transmission measurement of IR spectra of silver-coated BaF_2 plates on which the perspiration was sampled.[103]

The observation of enhanced infrared spectra may provide us with information on the interaction of adhesive molecules with the metal surfaces and the events taking place at the metal/solid-electrolyte interface as well. This method is thus a promising tool for investigating trace species on solid surfaces. However, it is not adequate for the investigation of well defined metal surfaces, and the available metals are limited. In addition, the signal intensity changes notably depending upon the morphology of the metal film. Hence this technique requires the use of an internal standard for the quantitative analysis of surface species.

2.2. Surface Electromagnetic Wave Spectroscopy in the Infrared Region

The conduction electron gas in metals may be excited by the incidence of an electron to form a collective longitudinal oscillation of a volume plasmon. In addition to the volume plasmon, there exists in metals another plasmon localized near the surface and called the surface plasmon, whose frequency, ω_{sp}, is given by Eq. (20), when the metal is in contact with a medium of dielectric constant ε_0:

$$\omega_{sp} = \omega_p / \sqrt{1 + \varepsilon_0} \qquad (21)$$

where ω_p ($\equiv \sqrt{4\pi n_0 e^2 / m}$) is the volume plasma frequency, and n_0 and m are the electron concentration and the mass of an electron, respectively. The light propagating to the surface combines under appropriate conditions with this surface plasmon to produce a surface plasmon polariton (SPP), i.e., a surface electromagnetic wave (SEW).

The infrared SEW propagates a few centimeters along the smooth surface of a semi-infinite metal. Hence the infrared SEW can be used as a sensitive probe of metal surfaces, and adsorbed molecules or oxide films on the surfaces as well as the film thickness can be investigated with the infrared SEW.[104-109]

Surface electromagnetic waves is non-radiative and propagates along the metal surface with an exponentially decaying field on each side of the surface, as shown in Fig. 18. Then the wave(-number) vector of the SEW has components in the x- and z-directions. The x-component of the wave vector is given by[110].

$$k_x = \omega/c \{\varepsilon_0\varepsilon_1/(\varepsilon_0+\varepsilon_1)\}^{1/2} = k_x' - ik_x'' \tag{22}$$

where ε_0 and $\varepsilon_1(= \varepsilon_1' - i\varepsilon_1'')$ are the dielectric functions of the ambient and the metal substrate, respectively. The z-components of the wave vector of the SEW are given by[107]

$$k_{1z} = \{\varepsilon_1(\omega/c)^2 - k_x^2\}^{1/2} \tag{23}$$

$$k_{0z} = (\{\varepsilon_0 (\omega/c)^2 - k_x^2\}^{1/2} \tag{24}$$

where k_{0z} and k_{1z} are the z-components in the ambient and the metal, respectively. In order that the z-components of the wave vector decay exponentially

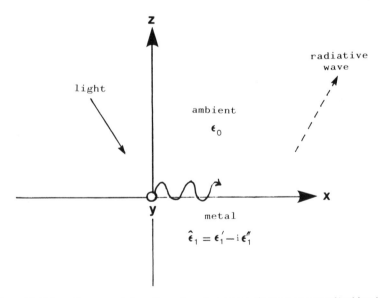

Figure 18. Schematic representation of a surface electromagnetic wave on a metal/ambient interface.

and are imaginary, we find from Eqs. (22) and (24) that (ignoring absorption)[107]

$$\varepsilon_0 + \varepsilon_1' < 0 \tag{25}$$

This is the condition for the SEW to be excited on the metal/ambient interface. The frequency for which $\varepsilon_0 + \varepsilon_1' = 0$ is the surface plasmon frequency ω_{sp} and corresponds to the frequency of SEW for $k_x' \rightarrow \infty$. This is the limiting frequency of the SEW and changes depending upon the metal employed. When the ambient is vacuum or air, the real part, k_x', can be expressed as in Eq. (26), because generally $|\varepsilon_1'| \gg \varepsilon_1''$ for free electron metals:

$$k_x' = \omega/c\{\varepsilon_1'/(\varepsilon_1' + 1)\} \tag{26}$$

Since $\varepsilon_1' < -1$ (or $-\varepsilon_0$), k_x' is always larger than ω/c, though nearly equal to ω/c for low frequencies wherein $\varepsilon_1' \ll -1$. The SEW dispersion curve therefore, lies below the photon line, as shown in Fig. 19. The imaginary part, k_x'' ($= 1/2L_x$), shows the decay of the SEW during the propagation, and the propagation distance L_x is defined as the distance required for the intensity of the SEW to decay to $1/e$ of its initial value.

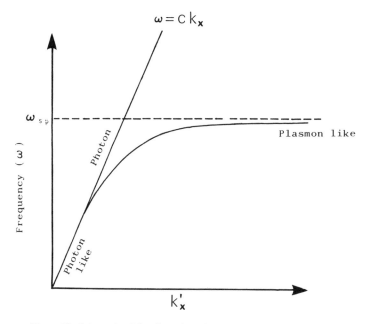

Figure 19. Schematic of the dispersion of a surface electromagnetic wave.

Figure 19 shows that the momentum of the SEWs is always larger than that of photons— in other words, the wavelength of an SEW is shorter than that of a photon of the same frequency traveling in vacuum. In consequence, the SEW cannot be excited directly by the incidence of light on a smooth metal surface.

Various methods have been discovered for exciting SEW. The deficiency in momentum of the photon can be offset by introducing a periodic structure on the metal surface.[111,112] When a plane wave of light is incident onto a metal grating, the x-component of the photon wave vector becomes

$$k_x = (\omega/c)\sin \phi_1 \pm N(2\pi/a) \qquad (27)$$

where a is the grating spacing, N the diffraction order, and ϕ_1 is the angle of incidence. This equation shows that the x-component of the wave vector of the incident radiation may become equal to k_x' of the SEW at appropriate ϕ_1 and a. Hence, the SEW can be excited by choosing the angle of incidence of light on the metal surface with a grating of spacing a. The SEW thus excited can be converted into the photon wave of the same frequency by another grating, as shown in Fig. 20a.

Two prisms can be used instead of the gratings to launch and recover the SEW (Fig. 20b).[113] The light beam is incident near the corner of the prism so as to minimize the recoupling of the SEW with the prism and maintain the SEW signal as high as possible.

In these two configurations, the SEW travels a fairly large distance, and the number of the surface species which are sampled increases. The absorption of the surface species could be measured at a high sensitivity by measuring the intensity of the recovered radiation or by varying the distance between the two prisms and measuring the change in propagation distance due to the absorption of the surface species. However, the throughput of both the arrangements is low and a highly collimated beam is required. Infrared lasers therefore, have been employed for exciting the SEWs, and the measurements are made only at several discrete frequencies.

For improving the low throughput, the edge coupling configuration (Fig. 20c) was invented by Chabal and Sievers.[114] In this arrangement, a thin metal film (~1 μm) is evaporated on the prism base and a laser beam is incident on the prism base at an appropriate angle for the coupling. Schlesinger and Sievers[115] and later Chesters et al.[116] succeeded in launching and recovering broad band SEWs, using the edge coupling configuration.

The single prism configuration (Fig. 9) can be used for the excitation and detection of SEWs. In an optically dense medium having a refractive index, n_p, the traveling light has a wavelength λ_0/n_p, where λ_0 is the wavelength in

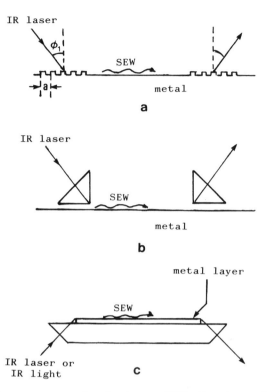

Figure 20. Various arrangements to launch and recover SEWs: (a) grating coupling, (b) two-prism coupling, and (c) edge coupler.

vacuum. Then the x-component of the wave vector of the light in the prism is $n_p k_{p0} \sin \phi_1$, where k_{p0} $(= \omega/c)$ is the wave vector of the light in vacuum. The incident angle ϕ_1 must be larger than the critical angle. The thickness, t, of the gap between the prism base and the metal surface should be small enough to generate the SEW by the evanescent wave of the totally reflecting radiation but large enough to minimize recoupling of the SEW back into the prism. This configuration (Fig. 9a) was invented by Otto[117] and bears his name. The gap may be replaced with an optically rare and transparent material.

When a thin metal layer is deposited on the prism base, the layer serves as the gap and the SEW is excited on the metal/ambient interface (Fig. 9b).[118] This configuration is called the Kretschmann configuration. Ueba and Ichimura[119] and Ishida *et al.*[67,120] have computed the electric field strength normal to the surface of Ag films on CaF_2 prisms, using the multilayer Fresnel formula, because the SEW accompanies the field predominantly normal to the metal surface. Although the results obtained are different in detail from one another, both calculations indicate that a highly enhanced field is obtained in the Kretschmann configuration by the incidence of p-polarized light at an angle near the critical angel. The maximum field enhancement could be obtained by the use of rather thin silver films because silver absorbs strongly the infrared radiation, and the optimum film thickness was calculated to be ~30 nm at 3000 cm^{-1}.[119] These results do not agree with the observations described in the preceding section, because no SEW was generated on the evaporated thin films.

A corresponding calculation for the Otto configuration shows that the strongest electric field is obtained by the incidence of p-polarized light at an angle near the critical angle, and the optimum angle of incidence decreases with increasing wavelength.[120] The absorption intensity of the sample measured with the enhanced field of the SEW increases with the thickness of the sample film while the thickness is very small, but begins to decrease because of the damping of the SEW when the film thickness increases beyond a pivotal value, e.g., ~ 10 nm for the carbonyl vibration band of PVAc. On the other hand, the absorption enhancement shows a strong frequency dependence, resulting in the remarkable change of the relative intensity of the absorption bands. The optimum gap thickness was also investigated in this work.

In the Otto configuration, the incidence of the IR beam at the angle somewhat lower than the critical angle brings about an increase in the electric field strength on the metal surface arising from the interference of the traveling wave.[120,121] Although the enhancement of electric field is much smaller than that in the SEW, enhanced absorption spectra can be measured without noticeable distortion by the use of the enhanced field.

Ishino and Ishida utilized a multiple reflection modification of the Otto configuration in the observation of the SEW spectra of PVAc films on copper substrates.[122] Figure 21 shows the SEW spectrum of a film of 5 nm in thickness obtained with 25 reflections. The external reflection spectrum of the same sample obtained with a single reflection arrangement is also shown in this figure. This figure demonstrates that multiple reflection SEW spectroscopy is a feasible technique for the observation of thin films on metal substrates.

Hjortsberg et al. measured SEW spectra of Mn stearate films of 2 and 10 monolayers with the Otto configuration.[123] Lange et al. employed the excitation of the SEW in the Kretschmann configuration combined with the potential modulation method for the investigation of species in the Helmholtz layer at the sulfuric acid/Au electrode interface.[124] Since the SEW decays rapidly in the wavelength region where the solvent is absorbing, the O–H stretching vibration

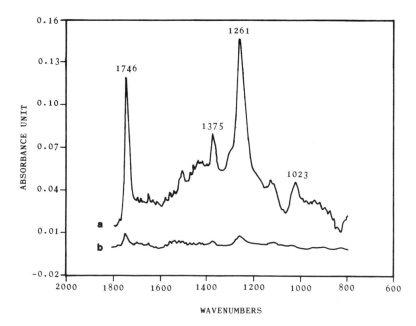

Figure 21. (a) SEW spectrum of a 5-nm PVAc film on copper measured with 25 reflections. (b) Single external reflection spectrum of the same sample as (a): p-polarized IR beam was incident in both the measurements [Reprinted with permission from Y. Ishino, and H. Ishida, *Anal. Chem.* **58**, 2448 (1986). © 1986 American Chemical Society].

bands were investigated using D_2O as the solvent. The decrease in intensity and the shift of absorption peak of the absorption bands were seen upon the shift of electrode potential to both positive and negative values from the potential of zero charge.

The application of SEW spectroscopy to the study of metal/electrolyte interfaces was later performed by Schellenberger *et al.*[125] Since the SEW is strongly damped by water, the emmersed electrode technique[126] was used in the measurement, and the underpotential deposition of Pb on a Ag substrate was studied employing a cw–CO_2 laser ($\lambda = 10.6$ µm) as the light source. A notable decrease in the intensity of the SEW was detected upon deposition of the Pb overlayer. Surface electromagnetic wave spectroscopy was also used for the measurement of infrared spectra of physisorbed benzene on copper[127] and hydrogen adsorbed on W[100].[128]

References

1. N. J. Harrick, *J. Opt. Soc. Amer.* **49,** 376 (1959).
2. J. Fahrenfort, *Spectrochim. Acta* **17,** 698 (1961).
3. N. J. Harrick, *Internal Reflection Spectroscopy,* Interscience, New York (1967); Harrick Scientific Corp., Ossining, NY (1979).
4. F. M. Mirabella and N. J. Harrick, *Internal Reflection Spectroscopy, Review and Supplement,* Harrick Scientific Corp., Ossining, NY (1985).
5. W. N. Hansen, in: *Advances in Electrochemistry and Electrochemical Engineering,* Vol. 9, *Optical Techniques in Electrochemistry* (R. H. Muller, ed.), Wiley-Interscience, New York (1973), p. 1.
6. N. J. Harrick and K. H. Beckmann, in *Characterization of Solid Surfaces* (P. F. Kane and G. B. Larabee, eds.), Plenum, New York (1974), p. 215.
7. W. W. Wendlandt and H. G. Hecht, *Reflectance Spectroscopy,* Wiley-Interscience, New York (1966), Chap. 6.
8. V. A. Burrows, *Solid-State Electron.* **35,** 231 (1992)
9. K. Ohta and R. Iwamoto, *Appl. Spectrosc.* **39,** 418 (1985).
10. N. J. Harrick and F. K. du Pré, *Appl. Opt.* **5,** 1739 (1966).
11. J. E. Olsen and F. Shimura, *Appl. Phys. Lett.* **53,** 1934 (1988).
12. J. S. Wong, A. J. Rein, D. Wilks, and P. Wilks, Jr., *Appl. Spectrosc.* **38,** 32 (1984).
13. A. V. Annapragada, K. F. Jensen, and T. E. Keuch, *Nicholet FT-IR Spectral Lines* **12,** 2 (1991).
14. P. A. Flournoy and W. J. Schaffers, *Spectrochim. Acta* **22,** 5, 15 (1966).
15. C. S. Paik Sung and J. P. Hobbs, *Chem. Eng. Comm.* **30,** 229 (1984).
16. V. N. Zolotarev, V. A. Veremei, and T. A. Gorbunova, *Opt. Spectrosc.* **31,** 40 (1971).
17. G. E. Becker and G. W. Gobeli, *J. Chem. Phys.* **38,** 2942 (1963).
18. R. Butz, E. M. Oellig, H. Ibach, and H. Wagner, *Surf. Sci.* **147,** 343 (1984).
19. H. Froitzheim, U. Köhler, and H. Lammering, *Surf. Sci.* **149,** 537 (1985).
20. Y. J. Chabal, *Surf. Sci.* **168,** 594 (1986).
21. G. Lu and J. Crowell, *J. Chem. Phys.* **98,** 3415 (1993).
22. E. Yablonovitch, D. L. Allara, C. C. Chang, T. Gmitter, and T. B. Bright, *Phys. Rev. Lett.* **57,** 249 (1986).
23. Y. J. Chabal, *Physica* **B170,** 447 (1991).
24. V. A. Burrows, Y. J. Chabal, G. S. Higashi, K. Raghavachari, and S. B. Christman, *Appl. Phys. Lett.* **53,** 998 (1988).
25. Y. J. Chabal, G. S. Highashi, K. Raghavachari, and V. A. Burrows, *J. Vac. Sci. Technol.* **A7,** 2104 (1989).
26. T. Takahagi, I. Nagai, A. Ishitani, H. Kuroda, and Y. Nagasawa, *J. Appl. Phys.* **64,** 3516 (1988).
27. G. S. Higashi, Y. J. Chabal, G. W. Trucks, and K. Raghavachari, *App. Phys. Lett.* **56,** 656 (1990).
28. P. Jakob and Y. J. Chabal, *J. Chem. Phys.* **95,** 2897 (1991).
29. P. O. Hahn and M. Henzler, *J. Appl. Phys.* **52,** 4122 (1981).
30. A. Ogura, *J. Electrochem. Soc.* **138,** 807 (1991).
31. P. Jakob, P. Dumas, and Y. J. Chabal, *Appl. Phys. Lett.* **59,** 2968 (1991).
32. P. Dumas, Y. J. Chabal, and P. Jakob, *Surf. Sci.* **269/270,** 867 (1992).

33. P. Dumas, Y. J. Chabal, and G. S. Higashi, *Phys. Rev. Lett.* **65**, 1124 (1990); *J. Electron Spectrosc. Relat. Phenom.* **54/55**, 103 (1990).
34. F. M. Hoffman, *Surf. Sci. Rep.* **3**, 107 (1983).
35. P. Jakob, Y. J. Chabal, and K. Raghavachari, *Chem. Phys. Lett.* **187**, 325 (1991).
36. Y. J. Chabal and C. K. N. Patel, *Phys. Rev. Lett.* **53**, 210 (1984).
37. Y. J. Chabal and C. K. N. Patel, *Rev. Mod. Phys.* **59**, 835 (1987).
38. K. Kern, Y. Chabal, G. HIgashi, A. vom Felde, and M. Cardillo, *Chem. Phys. Lett.* **168**, 203 (1990).
39. C. T. Lenczycki and V. A. Burrows, *Thin Solid Films* **193**, 610 (1990).
40. A. vom Felde, K. Kern, G. S. Higashi, Y. J. Chabal, S. B. Christman, C. C. Bahr, and M. J. Cardillo, *Phys. Rev.* **B42**, 5240 (1990); A. vom Felde, C. Bahr, K. Kern, G. S. Higashi, Y. J. Chabal, and M. J. Cardillo, *Phys. Rev.* **B42**, 6865 (1990).
41. K. H. Beckmann, *Surf. Sci.* **5**, 187 (1966).
42. V. A. Burrows and J. Yota, *Thin Solid Films* **193**, 371 (1990).
43. J. Yota and V. A. Burrows, *J. Appl. Phys.* **69**, 7369 (1991).
44. Y. Nagasawa, H. Ishida, F. Soeda, A. Ishitani, I. Yoshii, and K. Yamamoto, *Mikrochim. Acta* **8**, 431 (1988); Y. Nagasawa, I. Yoshii, K. Naruke, K. Yamamoto, H. Ishida, and A. Ishitani, *J. Appl. Phys.* **68**, 1429 (1990).
45. G. L. Haller and R. W. Rice, *J. Phys. Chem.* **74**, 4386 (1970).
46. R. Maoz and J. Sagiv, *J. Colloid Interface Sci.* **100**, 465 (1984).
47. T. Takenaka, K. Nogami, H. Gotoh, and R. Gotoh, *J. Colloid Interface Sci.* **35**, 395 (1971).
48. F. Kimura, J. Umemura, and T. Takanaka, *Langmuir* **2**, 96 (1986).
49. J. P. Hobbs, C. S. Paik Sung, K. Krishnan, and S. Hill, *Macromolecules* **16**, 193 (1983).
50. A. Garton, D. J. Carlsson, and D. M. Wiles, *Appl. Spectrosc.* **35**, 432 (1981).
51. L. Wang and R. S. Porter, *J. Appl. Polymer Sci.* **28**, 1439 (1983).
52. P. V. Ashrit, S. Badilescu, F. E. Girouard, and V.-V. Truong, *Appl. Opt.* **28**, 420 (1989).
53. M. I. Tejedor-Tejedor and M. A. Anderson, *Langmuir* **2**, 203 (1986).
54. M. C. Pham, F. Adami, P. C. Lacaze, and J. C. Dubois, *J. Electroanal. Chem.* **201**, 413 (1986); *J. Electroanal. Chem.* **210**, 295 (1986).
55. D. J. Carlsson and D. M. Wiles, *Macromolecules* **4**, 174 (1971).
56. P. Blais, D. J. Carlsson, and D. M. Wiles, *J. Polymer Sci.* A-1, **10**, 1077 (1972).
57. D. Briggs, V. J. I. Zichy, D. M. Brewis, J. Comyn, R. H. Dahn, M. A. Groon, and M. B. Konieczko, *Surf. Interface Anal.* **2**, 107 (1980).
58. R. Iwamoto and K. Ohta, *Appl. Spectrosc.* **38**, 359 (1984).
59. C. Sellitti, J. L. Koenig, and H. Ishida, *Carbon* **28**, 221 (1990).
60. D. G. Cameron, D. Escolar, T. G. Goplan, A. Nadeau, R. P. Young, and R. N. Jones, *Appl. Spectrosc.* **34**, 646 (1980).
61. C. S. Blackwell, P. J. Degen, and F. D. Osterholtz, *Appl. Spectrosc.* **32**, 480 (1978).
62. M. J. Dignam and S. Mamiche-Afara, *Spectrochim. Acta* **44A**, 1435 (1988).
63. J. E. Bertie and H. H. Eysel, *Appl. Spectrosc.* **39**, 392 (1985).
64. J. B. Huang and M. W. Urban, *Appl. Spectrosc.* **46**, 1666 (1992).
65. W. Suëtaka, *J. Spectrosc. Soc. Jpn.* **17**, 231 (1968).
66. Y. Ishino and H. Ishida, *Appl. Spectrosc.* **42**, 1296 (1988).
67. K. Ohta and H. Ishida, *Appl. Opt.* **29**, 1952 (1990).
68. S. Badilescu, P. V. Ashrit, F. E. Girouard, and V.-V. Truong, *J. Electrochem. Soc.* **136**, 3599 (1989).
69. C. Sellitti, J. L. Koenig, and H. Ishida, *Appl. Spectrosc.* **44**, 830 (1990).

70. A. Hartstein, J. R. Kirtley, and J. C. Tsang, *Phys. Rev. Lett.* **45**, 201 (1980).
71. T. Kamata, A. Kato, J. Umemura, and T. Takenaka, *Langmuir* **3**, 1150 (1987).
72. S. Badilescu, P. V. Ashrit, and V.-V. Truong, *Appl. Phys. Lett.* **52**, 1551 (1988).
73. Y. Suzuki, Y. Terui, A. Hatta, and W. Suëtaka, *J. Spectrosc. Soc. Jpn.* **41**, 164 (1992).
74. A. A. Sigarev and V. A. Yakovlev, *Opt. Spectrosc.* (USSR) **56**, 336 (1984).
75. A. Hatta, T. Ohshima, and W. Suëtaka, *Appl. Phys.* **A29**, 71 (1982).
76. Y. Suzuki, M. Osawa, A. Hatta, and W. Suëtaka, *Appl. Surf. Sci.* **33/34** 875 (1988).
77. A. Hatta, N. Suzuki, Y. Suzuki, and W. Suëtaka, *Appl. Surf. Sci.* **37**, 299 (1989).
78. A. Hatta, Y. Suzuki, and W. Suëtaka, *Appl. Phys.* **A35**, 135 (1984).
79. S. Badilescu, P. V. Ashrit, V.-V. Truoung, and I. I. Badilescu, *Appl. Spectrosc.* **43**, 549 (1989).
80. D. A. G. Bruggeman, *Annal. Phys. (Leipzig)* **24**, 636 (1935).
81. A. Hatta, Y. Suzuki, T. Wadayama, and W. Suëtaka, *Appl. Surf. Sci.* **48/49**, 222 (1991).
82. Y. Nishikawa, K. Fujiwara, and T. Shima, *Appl. Spectrosc.* **44**, 661 (1990).
83. J. P. Devlin and K. Consani, *J. Phys. Chem.* **85**, 2597 (1981).
84. T. Wadayama, Y. Momota, A. Hatta, and W. Suëtaka, *J. Electroanal. Chem.* **289**, 29 (1990).
85. S. Pons, S. B. Khoo, A. Bewick, M. Datta, J. J. Smith, A. S. Hinman, and G. Zachmann, *J. Phys. Chem.* **88**, 3575 (1984).
86. T. Wadayama, T. Sakurai, S. Ichikawa, and W. Suëtaka, *Surf. Sci.* **198**, L359 (1988).
87. D. S. Corrigan, J. F. Foley, P. Gao, S. Pons, and M. J. Weaver, *Langmuir* **1**, 616 (1985).
88. M. Osawa and M. Ikeda, *J. Phys. Chem.* **95**, 9914 (1991).
89. M. Osawa and K. Ataka, *Surf. Sci.* **262**, L118 (1992).
90. M. Osawa, K. Ataka, K. Yoshii, and Y. Nishikawa, *Appl. Spectrosc.* **47**, 1497 (1993).
91. N. Emeric and A. Emeric, *Thin Solid Films* **1**, 13 (1967).
92. H. Shopper, *Z. Phys.* **130**, 564 (1951).
93. S. Yoshida, *Oyo Butsuri* **41**, 324 (1972).
94. Y. Nakao and H. Yamada, *Surf. Sci.* **176**, 578 (1986); *J. Electron Spectrosc. Relat. Phenom.* **45**, 189 (1987).
95. A. Hatta, Y. Chiba, and W. Suëtaka, *Surf. Sci.* **158**, 616 (1985); *Appl. Surf. Sci.* **25**, 327 (1986).
96. A. Hatta, Y. Sasaki, and W. Suëtaka, *J. Electroanal. Chem.* **215**, 93 (1986).
97. M. Osawa, K. Ataka, K. Yoshii, and T. Yotsuyanagi, *J. Electron Spectrosc. Relat. Phenom.* **64/65**, 371 (1993).
98. K. Kunimatsu, H. Seki, and W. G. Golden, *Chem. Phys. Lett.* **108**, 195 (1984).
99. K. Kunimatsu, H. Seki, W. G. Golden, J. G. Gordon, II, and M. R. Philpott, *Surf. Sci.* **158**, 596 (1985).
100. K. Nakamoto, *Infrared and Raman Spectra of Inorganic and Coordination Compounds,* 4th ed., Wiley-Interscience, New York (1986), p. 276.
101. N. Matsuda, K. Yoshii, K. Ataka, M. Osawa, T. Matsue, and I. Uchida, *Chem. Lett.* **1992**, 1385 (1992).
102. Y. Nishikawa, K. Fujiwara, K. Ataka, and M. Osawa, *Anal. Chem.* **65**, 556 (1993).
103. Y. Nishikawa, K. Fujiwara, M. Osawa, and K. Takamura, *Anal. Sci.* **9**, 811 (1993).
104. V. M. Aganovich and D. L. Mills, eds., *Surface Polarizations,* North Holland, Amsterdam (1982).
105. R. J. Bell, R. W. Alexander, Jr., and C. A. Ward, in: *Vibrational Spectroscopies for Adsorbed Species* (A. T. Bell and M. L. Hair, eds.), American Chemical Society, Washington, D.C. (1980), p. 99.
106. Y. J. Chabal, *Surf. Sci. Rep.* **8**, 211 (1988).

107. R. J. Bell, R. W. Alexander, Jr., C. A. Ward, and I. L. Tyler, *Surf. Sci.* **48**, 253 (1975).
108. G. N. Zhizhin, E. A. Vinogradov, M. A. Moskalova, and V. A. Yokovlev, *Appl. Spectrosc. Rev.* **18**, 171 (1982).
109. Z. Lenac and M. S. Tomaš, *Surf. Sci.* **154**, 639 (1985).
110. R. J. Bell, R. W. Alexander, Jr., W. F. Parks, and G. Kovener, *Opt. Comm.* **8**, 147 (1973).
111. U. Fano, *J. Opt. Soc. Am.* **31**, 213 (1941).
112. R. H. Ritchie, E. T. Arakawa, J. J. Cowan, and R. N. Hamm, *Phys. Rev. Lett.* **21**, 1530 (1968).
113. J. Schoenwald, E. Burstein, and J. M. Elson, *Solid State Comm.* **12**, 185 (1973).
114. Y. J. Chabal and A. J. Sievers, *Appl. Phys. Lett.* **32**, 90 (1978).
115. Z. Schlesinger and A. J. Sievers, *Surf. Sci.* **102**, L29 (1981).
116. M. A. Chesters, S. F. Parker, and V. A. Yakovlev, *Opt. Commun.* **55**, 17 (1985).
117. A. Otto, *Z. Phys.* **216**, 398 (1968).
118. E. Kretschmann and H. Raether, *Z. Naturforsch.* **A23**, 2135 (1968).
119. H. Ueba and S. Ichimura, *Surf. Sci.* **118**, L273 (1982).
120. Y. Ishino and H. Ishida, *Surf. Sci.* **230**, 299 (1990).
121. Y. Suzuki, S. Shimada, A. Hatta, and W. Suëtaka, *Surf. Sci.* **219**, L575 (1989).
122. Y. Ishino and H. Ishida, *Anal. Chem.* **58**, 2448 (1986).
123. A. Hjortsberg, W. P. Chen, E. Burstein, and M. Pomerantz, *Opt. Comm.* **25**, 65 (1978).
124. P. Lange, V. Glaw, H. Neff, E. Plitz, and J. K. Sass, *Vacuum* **33**, 763 (1983).
125. U. Schellenberger, M. Abraham, and U. Trutschel, *Surf. Sci.* **192**, 555 (1987).
126. W. N. Hansen, *Surf. Sci.* **101**, 109 (1980).
127. K. Bhasin, D. Bryan, R. W. Alexander, Jr., R. J. Bell, *J. Chem. Phys.* **64**, 5019 (1976).
128. Y. J. Chabal and A. J. Sievers, *Phys. Rev. Lett.* **44**, 944 (1980).

Infrared Emission
Spectroscopy

1. Introduction

A molecule in a vibrationally excited state has a certain probability of emitting infrared radiation in the presence or in the absence of incident electromagnetic radiation, resulting in induced and spontaneous emission, respectively. At room temperature ($T \simeq 300K$), the number of molecules in a first excited state is less than 1% of the population in the ground state, when the separation of energy levels is about 1000 cm^{-1}, typical in the infrared. The population should be negligible in still higher states. In consequence, the induced infrared emission is much weaker than the (induced) absorption. However, the spontaneous infrared emission is stronger than the induced emission under the experimental conditions generally employed in observing surface species, as described in Section 2.1 of Chapter 2. The infrared emission spectrum of a thin film on a solid surface, therefore, may easily be observed at moderately elevated temperatures.

Infrared spectra of thin films on metal substrates can be obtained by the oblique incidence reflection method described in Chapter 2. However, when the temperature of the sample is raised, the emission spectrum is superimposed on the absorption spectrum, resulting in the distortion of absorption bands. Although the application of the polarization modulation technique may reduce the distortion, as described in Section 2.1 of Chapter 2, a rise in sample temperature beyond 500 K makes it difficult to observe reflection spectra in the long wavelength region. In addition, the oblique incidence reflection method is not quite feasible for observing a thin film on a rough or a curved metal surface. For this reason, the emission measurement should be useful in observing the foregoing samples. In this chapter, the general features of infrared emission from species on bulk solid surfaces will be discussed. Typical applications of emission IR spectroscopy to the study of thin films and adsorbed species on bulk solid substrates will be explored in some detail.

Generally speaking, infrared emission spectroscopy has been employed in the investigation of fused salts and silicates, flame and catalytic reactions besides the thin films, and adsorbed species on bulk solid substrates, and some review articles about IR emission spectroscopy as a whole have been published.[1,2] For example, Sheppard has demonstrated the advantages of FTIR spectroscopy in the observation of IR emission spectra from samples at moderately elevated temperatures. He recommended the measurement of the ratio between the emission spectrum of the sample and the blackbody reference at the same temperature.[2]

2. Optimum Experimental Conditions

After the exploratory experiments of Eischens and Pliskin,[3] Low and Inoue showed that discernible IR emission bands of films on metal surfaces could be obtained at high temperatures employing a dispersive spectrometer.[4] By using of a Fourier transform spectrometer, they later obtained emission spectra of organic compounds on metal surfaces at temperatures slightly above ambient.[5] The effect of thickness of the sample layer was investigated by Griffiths[6] and Fabbri and Balardi.[7] These works, where the emission normal to the metal surface was observed, showed that, whereas good spectra could be obtained when the sample was moderately thin, spectra from thicker layers did not retain the similarity to the absorption spectra because of self-absorption. Observations of normal emission were not successful in investigating very thin films or adsorbed species on metal surfaces. The dependence of the emitted intensity on the angle of observation was investigated experimentally by Blanke *et al.* by employing a dispersive spectrometer.[8] The emission spectrum from a film of 17 layers of stearate on a gold substrate increased in intensity with the angle of observation, as can be seen in Fig. 1. The signal-to-noise ratio can be enhanced in a multiple-reflection sampling system if the number of reflections remains restricted.[9] Emission spectra from an oxide layer of 6 nm thickness on aluminum could be recorded at high observation angles by means of an FTIR spectrometer.[10] An increase in emission intensity was also observed by increasing the angle of observation in this work.

The angular intensity distribution of the light emitted from a species adsorbed on a metal surface was investigated theoretically by Greenler.[11] He

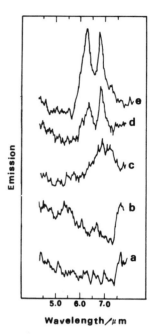

Figure 1. Emission spectra of a stearate film on gold at 375–377 K. Angle of observation: (a) 0°, (b) 30°, (c) 60°, (d) 75°, and (e) 85° [Reprinted from *Spectrochim. Acta*, **32A**, J. F. Blanke, S. E. Vincent, and J. Overend, Infrared spectroscopy of surface species: instrumental considerations, p. 170, © 1976, with kind permission from Pergamon Press Ltd., Headington Hill Hall, Oxford OX3 OBW, UK].

treated the infrared emission as well as the visible emission and showed that an intensity peak should appear at a high angle of observation in the infrared emission. The theory of the angular intensity distribution of infrared emission is now outlined along his treatment.

It is assumed that the infrared radiation is dipole radiation. Let us consider oscillating unit dipoles located in close proximity to the metal surface and oriented in three mutually perpendicular orientations, as shown in Fig. 2. Suppose that the radiation arriving at the detector is the sum of the direct wave (A_1) and the wave (A_2) that, after the reflection at the metal surface, travels in the same direction as A_1. Since the dipoles are separated at a distance negligible in comparison to the wavelength of radiated light, the optical path difference between A_1 and A_2 can be neglected. Writing A_0 for the amplitude of the wave radiated at right angle to the dipole axis, the resultant amplitude, A, after combination of the two waves is given by

$$A = A_0 L\{\exp(i\omega t) + |r_v|\exp i(\omega t + \beta + \gamma_v)\} \tag{1}$$

where $L = A_1/A_0 = A_2/A_0$ and β is an effective phase shift arising from the difference in the vector direction of the two waves. A_1 and A_2 are the amplitude

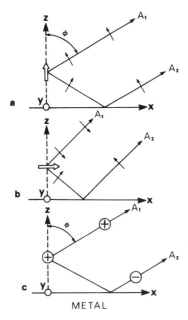

Figure 2. Emission from a dipole present in the immediate neighborhood of a metal surface. (a) Dipole oscillating normal to the metal surface, (b) dipole oscillating parallel to the surface and the plane of incidence, and (c) dipole oscillating parallel to the surface and perpendicular to the plane of incidence. ϕ = angle of observation [Reprinted from R. G. Greenler, *Surf. Sci.* **69**, 647 (1977)].

of the direct and the reflected rays, respectively. The Fresnel reflection coefficient at the metal surface, r_v, can be written in the form

$$r_v = |r_v| \exp(i\gamma_v), \quad v = p,s \tag{2}$$

where γ_v is the phase change on reflection for v-polarized light. p and s represent the light polarized parallel and perpendicular to the plane of the surface normal and the emitted light, respectively. The plane will be called the plane of incidence in this chapter, because it is the plane of incidence for the reflecting beam.

The time average of $|A|^2$ gives the intensity of the resultant light. Using Eq. (1) and letting the coefficient be unity, the intensity I can be written as

$$I = L^2\{1 + |r_v|^2 + 2|r_v|\cos(\beta + \gamma_v)\} \tag{3}$$

Let us now consider the three cases shown in Fig. 2. In case (a), where $L = \sin\phi$ and $\beta = 0$, Eq. (3) reduces to

$$I = \sin^2\phi\{1 + |r_p|^2 + 2|r_p|\cos\gamma_p\} \tag{4}$$

Since the reflectivity of noble metals is high at normal incidence in the infrared,

$|r_p|$ can be approximated at unity. At the grazing angle ($\phi = 90°$) incidence, it takes a value of unity. Consequently, when the incident angle is $0°$ or $90°$, Eq. (4) can be simplified to

$$I = 2\sin^2 \phi (1 + \cos \gamma_p) \qquad (5)$$

As described in Section 1.1 of Chapter 2, the phase shift, γ_P, is equal to $-\pi$ when $\phi = 90°$. So we have $I = 0$. This is also the case when $\phi = 0°$.

The solid line in Fig. 3 shows the angle dependence of emission intensity at a wavelength of 5 μm calculated by Eq. (4) for a dipole on a gold surface. This figure shows that the intensity of the infrared emission should reach a maximum value at a high observation angle near $80°$. The calculation shows that the maximum intensity is more than three times larger than that emitted

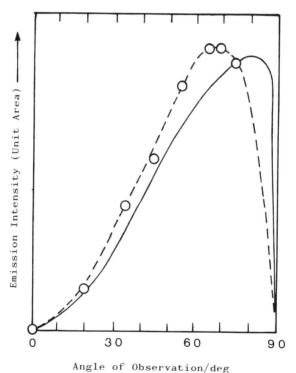

Figure 3. Angular distribution of emission intensity per unit area. ◯, observed intensity of the 630 cm^{-1} band of a Cu_2O thin film on gold; ——, calculated by Eq. (4); – –, calculated by Eq. (7) (adapted from K. Wagatsuma and W. Suëtaka, unpublished data).

from an isolated dipole ($|A_0|^2$). The increase in observation angle accompanies the increase in sampled surface area. If the normally oriented dipoles are uniformly distributed on the surface without mutual interaction, the measured intensity should increase in proportion to secϕ. In consequence, the measured emission intensity reaches a maximum at an angle larger than 80°. The increase in observed intensity reaches 30 times at an angle of observation of 85° in the above-mentioned condition.

In case (b), where $L = \cos \phi$, and $\beta = \pi$, Eq. (3) becomes

$$I = \cos^2 \phi[1 + |r_p|^2 + 2|r_p|\cos (\pi + \gamma_p)] \tag{6}$$

The phase shift on reflection (γ_p) is very small unless the angle of observation approaches 90°. It is shown that $|r_p|$ is nearly equal to unity whenever the observation angle is smaller than 85°, as far as highly reflective metals are concerned.[12] It follows that the observed intensity is negligible if an angle smaller than 85° is used. When the angle exceeds 85°, $\cos^2 \phi$ has a very small value that falls to zero at 90°, and the terms in the square brackets give a small value. In this case, therefore, the emitted intensity is very weak irrespective of observation angle.

In case (c), where $L = 1$ and $\beta = 0$, the phase shift on reflection is close to $-\pi$, as shown in Section 1.1 of Chapter 2. The observed intensity (I) is also very weak in this case, because $|r_s|$ is nearly equal to unity for highly reflective metals.

In conclusion, the light detected should be attributed to the dipoles oscillating normal to the metal surface, and its intensity will change depending upon the observation angle as the solid line in Fig. 3. It follows that the light emitted from species adsorbed on a highly reflective metal surface should be polarized in the plane of incidence.

Figure 4 shows emission spectra of a Glyptal thin film of 150 nm in thickness on a mild steel observed at 423 K.[13] In this figure, solid lines show that the emission of the light polarized parallel to the plane of incidence (*p*-polarized light) increases in intensity with the angle of observation. On the contrary, no spectrum was observed in the observation of the light polarized perpendicular to the incident plane (*s*-polarized light) irrespective of observation angle. The observed results are in perfect agreement with the theoretical deduction made by Greenler, confirming that, in the observation of the emission of species on metal surfaces, the measured signal is attributable exclusively to the component of the oscillating dipole normal to the surface. It must be stressed that, in the infrared investigation of species on metal surfaces, only the information about the normal component of dynamic dipoles on the surface is obtained, not only in the reflection measurement but also in the emission

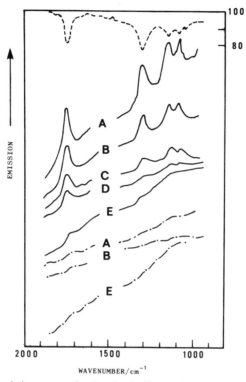

Figure 4. Infrared emission spectra of a glyptal resin film on mild steel. Film thickness = 150 nm, temperature = 423 K, Angle of observation, (A) 75°, (B) 60°, (C) 45°, (D) 30°, and (E) 0°. ——— , parallel polarization, − ·· − , perpendicular polarization, − − , reflection spectrum of the same sample at room temperature (angle of incidence = 70°) (adapted from Ref. 13).

observation. The surface selection rule, described in Section 1.1 of Chapter 2, holds for emission spectroscopy as well.

Figure 4 shows a spectrum of the same sample obtained with the oblique incidence reflection method for the purposes of comparison. The emission spectrum (A) has a comparable contrast to the reflection spectrum. The relative intensity of emission bands increases in the low frequency region. This increase results from the use of a high temperature infrared source as the reference in recording the emission spectra. Such a distortion can be removed by measuring the ratio between the emission spectrum of the sample against the blackbody emission at the same temperature. The improvement obtained by the ratio measurement will be described in Section 3 of this chapter.

The angular distribution and the polarization change of the emission intensity of the Glyptal film on the mild steel can be elucidated qualitatively by the theory mentioned above. Strictly speaking, however, the theory is applicable only to the emission of molecules adsorbed or present in a monomolecular layer on a highly reflective metal surface. In discussing the emission of a thicker film, one must take into account at least the self-absorption of light by the film and the optical path difference between the A_1 and the A_2 beams. Open circles in Fig. 3 show the observed emission intensity of the band at 630 cm^{-1} of a Cu_2O film (thickness = 160 nm) formed on a gold plate. The observed angular distribution of emission intensity does not agree with the solid line in the figure. This shows that Eq. (3) does not hold for the film of 160 nm in thickness. The open circles are in good agreement with the broken line in Fig. 3, which is the angular distribution calculated with the following approximation equation,[13] in which the self-absorption and the path difference are taken into account:

$$I = \exp(-\eta d)\sin^2 \phi[\{\exp(\eta d) - |r_p|^2\exp(-\eta d) - 1$$

$$+ |r_p|^2\}/\eta + 2|r_p|\{\sin(\xi d + \gamma_p) - \sin \gamma_p\}/\xi]$$

$$\eta = 4\pi k/\lambda\cos \phi, \; \xi = 4\pi(n - \sin^2 \phi)/\lambda\cos \phi \qquad (7)$$

where n and k are the refractive index and the extinction coefficient of the film at the temperature of measurement, d is the film thickness, and λ is the wavelength in vacuum.

In the preceding discussion, the emission of the substrate metal was neglected. When a small angle is used for the observation, the emission of highly reflective metals is very weak, ranging from 1 to 2 percent of the emission of the black body at the same temperature. When the angle of observation exceeds 80°, however, a sizable rise occurs in the intensity of the p-polarized emission followed by a sudden drop to zero at 90°. The intensity of s-polarized emission, on the contrary, remains very weak upon increasing the angle. The large angle emission of the metals thus, is polarized in the plane of incidence and is absorbed by the radiating surface species, generating an absorption spectrum of the very surface species. The absorption spectrum is superimposed on the emission spectrum of the surface species, reducing the contrast of the obtained spectrum.

Although the samples surface area increases with the angle of observation, as mentioned above, no appreciable intensity increase can be expected at high angles of observation because of the emission of the metal. In addition, the observation at high angles requires a very large surface area. We now have the optimum conditions for obtaining the emission spectrum of species on a metal surface: the p-polarized emission should be observed at a large angle in the range of 70°–80°.

With Eq. (7) we can estimate the change in emission intensity with the film thickness. When $d/\lambda \ll 1$, however, Eq. (7) can be expanded to terms of first order in ηd and ξd, except for near grazing angles of incidence. Retaining only first-order terms in (d/λ) of the obtained equation, we have (13)

$$I = d\sin^2 \phi \, (1 + |r_p|^2 + 2|r_p|\cos \gamma_p) \qquad (8)$$

Equation (8) shows that the intensity increases in proportion to the film thickness, when the film remains very thin.

Figure 5a shows the emission intensity of Cu_2O thin films of various thicknesses.[13] When the film thickness is less than 50 nm, the emission intensity increases in proportion to the film thickness in agreement with Eq. (8). When the thickness increases beyond 50 nm, the proportionality does not hold and the emission intensity increases in accordance with the solid line calculated with Eq. (7).[13] Figure 5b shows the intensity change of a moderately absorbing band of an acetylcellulose film.[14] The emission intensity increases in proportion to the film thickness until the thickness reaches 200 nm. The difference between these two figures can be attributed to the self-absorption. Since the band of Cu_2O is very strongly absorbing, the deviation from the proportionality begins at a small film thickness of 50 nm.

Now we pass to the emission from a thin film on a semiconductor such as silicon or germanium, which has a high refractive index and is transparent in the infrared region. The angular distribution of emission intensity from an oscillating dipole on a germanium plate was calculated with Eq. (3). The obtained result is shown in Fig. 6, which shows that a dipole oscillating perpendicular to the incident plane gives an emission of appreciable intensity, in contrast to the dipole on a metal surface. The dynamic dipole parallel to the incident plane still gives an emission stronger in intensity than the perpendicular dynamic dipoles and shows a maximum at a high angle of observation. It should be noted that the dynamic dipole in case (b) contributes to the parallel emission of a film on germanium. It is worthwhile to mention that the intensity maximum is observed at an angle larger than the 60° shown in Fig. 6, because the intensity per unit area is shown in this figure.

Figure 7 shows the angular distribution of intensity of the band arising from the antisymmetric vibration of the C–O–C group in a poly (vinyl acetate) film (thickness = 50 nm) on a silicon substrate. The maximum in intensity (per unit area) appears in this figure at an angle of 60°, in accordance with Fig. 6.

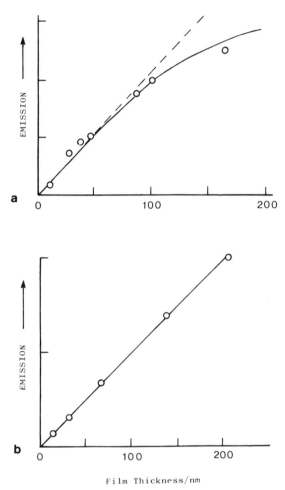

Figure 5. Dependence of emission intensity on film thickness. Observed band: (a) band at 630 cm^{-1} of Cu_2O, (b) band at 1365 cn^{-1} of acetylcellulose, O = Observed intensity, temperature = 433 K, angle of observation = 75° (adapted from Refs 13 and 14).

Comparing Fig. 7 with Fig. 6, the decrease in relative intensity is noticed at the maximum of emission intensity. This decrease is attributed to the self-absorption of the film, because the polymer film has an appreciable thickness.

The emission intensity of a film on a semiconductor surface increases with thickness in nearly the same manner as on a metal surface. Qualitatively

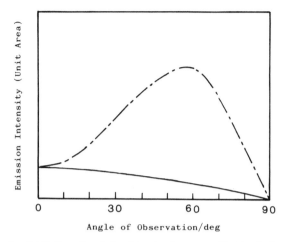

Figure 6. Angular distribution of emission intensity from an oscillating dipole on germanium. —— , Emission polarized perpendicular to the plane of incidence, — ·· — , emission polarized parallel to the plane of incidence (adapted from Ref. 14).

speaking, the emission characteristic of a film on a transparent substrate is in agreement with that of a film on a metal surface, but the emission of species on a metal surface is about ten times stronger at high observation angles than that from a molecule on a semiconductor surface.

The emission spectrum of a thin film on a metal or a semiconductor surface thus, can be obtained efficiently in the measurement of the emission at a high angle of observation. However, the sensitivity of this method is still insufficient for observing species adsorbed on a solid surface. An improvement in sensitivity therefore, is required for the observation of such species. The removal of background emission is efficient for the improvement, and spectrometers cooled entirely to a low temperature were employed for observing the monolayer or the adsorbed species on a metal surface. Employing a cryogenic interferometer operating at 77 K, Allara et al. observed an emission spectrum of a monolayer of p-nitrobenzoic acid on a copper substrate at 300 K.[15] Chiang et al. were successful in observing an emission spectrum of CO adsorbed on a clean nickel surface at room temperature by means of a newly designed liquid helium cooled grating spectrometer.[16,17] Although these cryogenic spectrometers are powerful for the observation of emission spectra of surface species at room temperature, they are too expensive to be used widely and a less expensive alternative is desired.

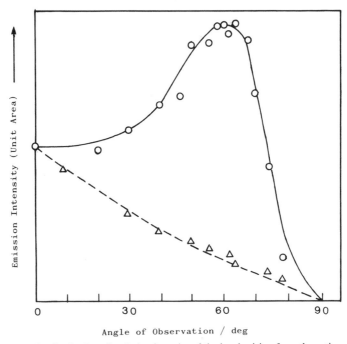

Figure 7. Angular distribution of emission intensity of the band arising from the antisymmetric stretching vibration of the C–O–C group of a poly(vinyl acetate) film on a silicon substrate. Film thickness = 50 nm, temperature = 423 K. —O—, observed intensity of p-polarized light, – Δ– , intensity of s-polarized light (adapted from T. Endo and W. Suëtaka, unpublished data).

Background emission can be minimized by the polarization modulation technique. The emission from a thin film on a metal surface is polarized parallel to the plane of incidence, as can be seen in Fig. 4. Figure 7 shows that most of the infrared light emitted from a thin film on a semiconductor is polarized parallel to the plane of incidence at high angles of observation. Consequently, the polarization modulation of the light emitted at high angles removes most of the background emission, bringing about an improvement in the S/N ratio.[18] It should be noted that polarization modulation cannot remove absorptions of gases present on the optical path, in sharp contrast to the reflection measurement, because it is the emission from the sample that induces the absorptions of gases and the intensity of absorption bands changes in accordance with the modulation frequency. An additional procedure is necessary to remove the absorptions.

For the analysis of emission spectra obtained from thick films, a rigorous theoretical treatment is necessary. Ohta and Ishida have calculated the light emissive power from multilayer films having temperature gradients by the use of matrix formulae.[19] The mixing of coherent and incoherent states is taken into account in the calculation. This method is useful for the analysis of the emission spectra from fairly thick films.

3. Instrumentation and Application of Infrared Emission Spectroscopy

In the measurement of infrared emission spectra of thin films on solid substrates, attention must be paid not only to the optimum conditions of observation but also to the absorption of the emitted light by gaseous molecules present on the optical path and the displacement of the emission peak frequency depending upon the sample temperature. Infrared emission measurements are generally performed at room temperature or moderately elevated temperatures. Since the emission peak appears in the infrared region at these temperatures, the observed relative intensity of the emission bands does not agree with the intensity in the absorption spectrum and remarkable background slopes are observed. The change in relative intensity may cause difficulty in the analysis of the obtained spectrum, because the molecular orientation gives rise to the change in relative intensity. The abnormal relative intensity and the background slopes can be corrected theoretically[15] or by dividing the emission spectrum of the sample by the emission from the substrate[16] or a blackbody.[10,20,21]

The light emitted from the sample is absorbed by water vapor and CO_2 present in the spectrometer as well as by ambient gases in the reaction chamber in which the sample is present. The former gases can be removed by evacuating the spectrometer. However, the presence of the latter gases is necessary for the *in situ* observation of the reaction at the solid surface. Furthermore, the emission from the metal substrate gives rise to an absorption spectrum of the gas, though it is weak in intensity. These absorptions of gas molecules may be removed by a ratio procedure.

Using an FTIR spectrometer, Kember *et al.* developed procedures for removing the atmospheric absorption and the background slope.[10,20] They obtained emittance spectra by measuring the ratio of the emission spectrum of the sample to the emission spectrum of a blackbody at the same temperature. The emission spectra thus obtained are closely related to absorption spectra.[20] However, the acquisition of a good interferogram of the weak emission requires

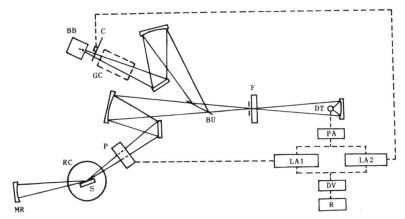

Figure 8. Experimental schematic of a double modulation infrared emission spectrometer. S, Sample, P, polarization modulator; C, mechanical chopper, BU, beam uniter; F, wavelength variable filters; BB, blackbody; DT, semiconductor detector; PA, pre-amplifier; R, recorder; LA, lock-in-amplifier; DV, analog divider; MR, spherical mirror; GC, gas-cell; RC, reaction chamber.

considerable time. The ratio at a short time interval may give a better emittance spectrum. A double modulation-double beam technique may permit taking the ratio at shorter time intervals.

Figure 8 is a schematic representation of a double beam infrared emission spectrometer, in which the radiation from the sample is modulated by a rotating polarizer (P) at a frequency of 65 Hz and is then incident upon a thin Ge beam-uniter (BU) at an incident angle of 76° (the Brewster angle of Ge) so as to minimize the loss in intensity of the light from the sample. A spherical mirror (MR) is used for effective collection of the radiation from the sample. At the same time, this mirror removes the stray light that is incident upon the sample mirror and goes to the detector after reflection. The radiation from a conical cavity (BB), used as the black-body, is modulated by a mechanical chopper (C) at a frequency of 2.3 kHz and reflected at the back surface of the beam-uniter. Both the radiation from the sample and the blackbody are monochromatized by wavelength-variable filters (F) and led to a liquid nitrogen cooled HgCdTe detector (DT). The modulation frequency of the rotating polarizer is too low to get the best performance of the detector. Although a photoelastic modulator can be used at a high modulation frequency, it causes too great a loss in intensity of the infrared light to be used in the spectrometer shown here.

The experimental results show that all of the background slope, the ab-

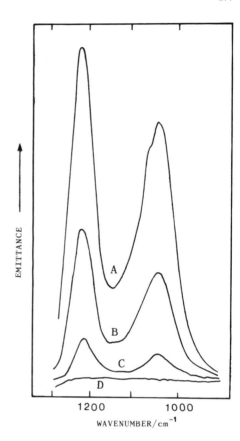

Figure 9. Emission spectra of thin acetylcellulose films on gold substrates. Each
spectrum is shifted vertically to avoid
overlapping. Temperature = 423 K, angle
of observation = 80°, film thickness: (A)
100 nm, (B) 50 nm, (C) 10 nm, and (D) 0
nm (bare gold) [Reprinted from T. Wadayama, W. Shiraishi, A. Hatta, and W.
Suëtaka, *Appl. Surf. Sci.* **44**, 43 (1990)].

sorption by the ambient gas, and the abnormal relative intensity of the emission
bands are removed by the double beam measurement. The absorption arising
from the reagent gas molecule may be removed by the introduction of a gas cell
into the optical path of the reference beam.

Figure 9 shows relative emittance spectra of acetylcellulose thin films on
gold substrates obtained with the double beam spectrometer shown in Fig. 8.
This figure shows that a good spectrum can be obtained from a thin film of 10
nm in thickness at 423 K and that the emission intensity increases in proportion
to the film thickness up to 100 nm.[21] The measurement of emittance spectra
has another advantage over the single beam measurement. While the emission
increases in intensity with temperature, the emittance of bands of the sample

remains constant independent of temperature. This is propitious for the estimation of film thickness.

Special attention should be paid to obtaining the emission spectrum of a thin film on a transparent substrate, such as germanium or silicon. When the sample is heated from behind, the metal oxide generally present on the heater surface emits a strong emission spectrum which passes through the substrate, often giving rise to an absorption spectrum of the sample film. To remove such awkward radiation, oxide-free metal such as gold or platinum is vacuum evaporated onto the back surface of the substrate or a thin plate of the metal is placed between the back surface and the heater. The evaporation of gold onto germanium or silicon is not desirable, however, because gold diffuses rapidly into the semiconductor at elevated temperatures. When the sample film has a thickness larger than 100 nm, the polarization modulation method does not provide a satisfactory emission spectrum, because the *s*-polarized emission has an intensity approaching that of the *p*-polarized light.

Tochigi *et al.* applied polarization modulation FTIR spectroscopy to the measurement of emission spectra from polymer films of ~10 nm thickness on metal substrates.[22] The emission from the sample was collected at an angle of 75° and modulated with a photoelastic modulator combined with a wire-grid

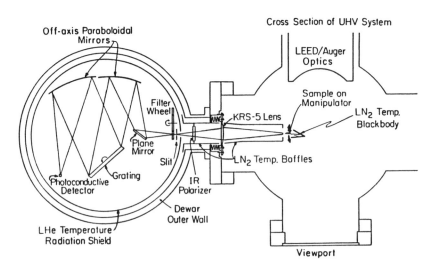

Figure 10. Optical layout of a cryogenic infrared emission apparatus [adapted from S. Chiang, R. G. Tobin, and P. L. Richards, *J. Electron Spectrosc. Relat. Phenom.* **29**, 113 (1983)].

polarizer. The step scan operation of the interferometer gives spectra of better signal-to-noise ratios than the continuous scan mode.

An example of cryogenic spectrometers is shown in Fig. 10.[16] This instrument consists of a liquid helium temperature grating spectrometer combined with an ultrahigh vacuum system. The sample is surrounded with buffers at liquid nitrogen temperature to minimize the background emission. Radiation from the sample, which is placed near grazing incidence, is focused on the

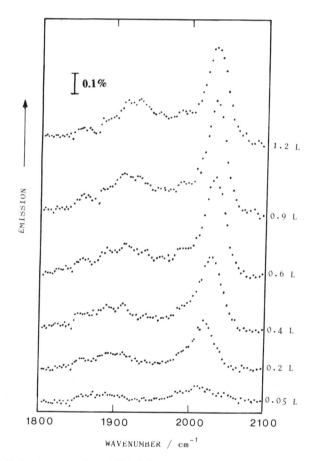

Figure 11. Emission spectra from Ni[100]8° stepped surface as a function of CO coverage [adapted from S. Chiang, R. G. Tobin, and P. L. Richards, *J. Electron Spectrosc. Relat. Phenom.* **29**, 113 (1983)].

entrance slit of the spectrometer by a KRS-5 lens. A cold polarizer is mounted for removing the radiation polarized perpendicular to the incident plane, which bears no information about surface species. A Si:Sb photoconductive detector is used for measuring signals from 330 cm^{-1} to 3000 cm^{-1}. Spectra of CO adsorbed on a clean Ni[100] 8° stepped surface were obtained with this instrument and are shown in Fig. 11. Two emission bands can be seen in the range of 2010–2035 cm^{-1} and 1850–1930 cm^{-1}. The former band is assigned to linearly adsorbed and the latter to bridge-bonded molecules.[16] On the other hand, no feature assignable to bridge-bonded CO appeared in the emission spectra of CO adsorbed on Ni[100] recorded at 310 K when the coverage was less than 0.5.[23] The CO–Ni mode was observed at 310 K for the same system.[24]

The CO–Pt vibrational mode was observed using a system of on-top CO on Pt[111] by emission spectroscopy in the temperature range of 210–400 K.[25] The changes in emission peak frequency, band shape, and band width were studied for changing coverage, and inhomogeneous broadening of the band was observed. The emission from the C–O stretching mode of the on-top species could be detected at a low CO coverage of 0.03. The emission spectra of bridge-bonded CO on a Pt[111] surface showed only one band of the C–O stretching vibration at 1849 cm^{-1}, in contrast to previous work.[26] The changes in bandwidth, peak frequency, and integrated intensity of this band were studied by changing the coverage and the temperature. From these observations, the "fault-line" model was deduced for high coverage structures. The temperature dependence of the bandwidth was attributed to the inhomogeneous broadening associated with an order-disorder transition in the overlayer.[27]

Figure 12 shows the emissivity spectrum of p-nitrobenzoic acid adsorbed on a native oxide-covered copper substrate obtained at 300 K with the cryogenic FTIR spectrometer.[15] The weak peak at ~1739 cm^{-1} is assignable to the carbonyl stretching vibration. A strong band at 1588 cm^{-1} was assigned to the antisymmetric stretching vibration of the carboxylate group; salt formation upon adsorption of the acid was deduced from these measurements.

Figure 13 shows the change in the emission spectrum of a thin poly(vinyl acetate) film on a silver substrate recorded at 523 K.[28] The intensity of emission bands decreases with time. The rate of decrease, however, changes depending upon the band. Whereas the bands at 1730 cm^{-1} (C=O stretching), 1375 cm^{-1} (CH$_3$ symmetric deformation), 1220 cm^{-1} (C–O–C antisymmetric stretching), and 1020 cm^{-1} (C–O–C symmetric stretching) decrease in intensity considerably, only a slight decrease is seen in the broad band around 1100 cm^{-1} (deformation of CH in the main chain). The difference in decreasing rate is a clue to elucidating the decomposition mechanism of the poly(vinyl acetate)

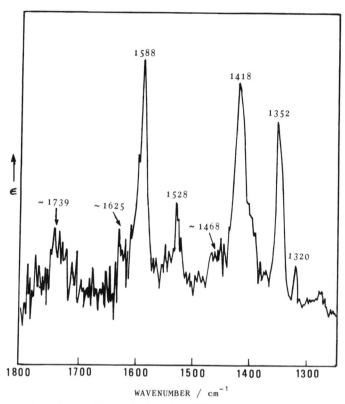

Figure 12. Spectrum of a monolayer *p*-nitrobenzoic acid film corrected for substrate emission and multiplied by the emissivity transform function. The ordinate is proportional to emissivity [adapted from D. L. Allara, D. Teicher, and J. F. Durana, *Chem. Phys. Lett.* **84**, 20 (1981)].

film. All of the emission bands exhibiting the extensive intensity decrease are attributed to the acetoxyl group of the side chain. This indicates that the thermal degradation proceeds through the break-off of the side chain.

Nearly the same rate of intensity decrease was observed in all of the emission bands of poly(methyl methacrylate) at 523 K.[28] This suggests that the breaking of the main chain takes place in the thermal decomposition of this compound. The carbonyl stretching band decreased in intensity at a slightly slower rate than the others. The slow decrease in emission intensity of the carbonyl group probably reflects the formation of this group during the decomposition. The thermal decomposition of poly(methyl methacrylate) in air is

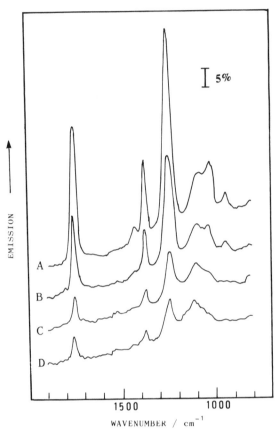

Figure 13. Change of emission spectrum of a poly(vinyl acetate) film observed *in situ* during thermal degradation. *p*-Polarized emission was measured. Temperature = 523 K, Angle of observation = 75°, Heating time: (A) 10 min, (B) 180 min, (C) 405 min, and (D) 520 min (adapted from Ref. 28).

accompanied by the oxidation of the polymer, producing the carbonyl groups.

Wadayama *et al.*[29] investigated the chemical vapor deposition of titanium oxide on gold substrates with the apparatus shown in Fig. 8. The growth of the oxide film was observed *in situ* at 523 K in a mixed gas of tetraisopropoxytitanium and oxygen. The absorption of gas-phase species was eliminated by the use of the gas cell shown in Fig. 8. This observation has revealed a modest growth rate in the film at the initial stage of deposition and the enhancement of film growth by the UV irradiation.

Mazzarese *et al.*[30] employed FTIR emission spectroscopy for monitoring at 773 K the behavior of the surface species on semiconductor substrates during organometallic vapor phase epitaxy. In this measurement, a gas flow cell was used. A GaAs substrate was placed on a gold plated copper mirror in the cell and the near normal emission from the sample was collected by a parabolic mirror. The above-mentioned signal enhancement due to the constructive interference with the reflected emission wave is hardly expected in this system, and the absorption of gas-phase species may be superimposed over the emission from the sample.

Trimethylgallium (TMGa) and ammonia were introduced in sequence (or in reverse order) in the gas cell, and the emission spectra were recorded after pumping down the gas phase to reduce the absorption of gaseous species. The obtained sample spectrum was divided by the background emission spectrum recorded separately. The emission bands of species formed on GaAs could be observed, but the surface species changed depending upon which species of TMGa or HN_3 was supplied first to the substrate surface. Though the emission IR measurement has given similar results to the ATR observations at lower temperatures, the decomposition of surface species at higher temperatures can be investigated only in the emission measurement.

Lauer and Vogel observed infrared emission spectra of deposits on metal strips in a fuel system simulator with an FTIR apparatus. The obtained spectra showed the presence of carboxylic acids, their salts, alkoxyl groups, and C=C bonds in the deposits. Species included in the deposits and their quantity varied with the addition of a small amount of heterocyclic compounds or metal naphthenates.[31] Low *et al.* applied infrared emission spectroscopy to the study of the mechanism of boundary lubrication. They observed emission spectra of oleic acid on a copper plate and found the formation of a binuclear copper oleate complex and its precursor. The diffusion of the free acid, its ester, and the complex on a copper surface was also studied.[32] Gratton *et al.* studied the oxidation of molybdenum.[33] The obtained spectra showed the formation of pseudocrystalline MoO_3 followed by the transition into the crystalline oxide. Matsui *et al.* studied the relation between the thickness of surface films and the emission intensity at 353 K using poly(vinyl acetate) films on aluminum plates as the samples.[34] They found that a linear correlation held between the sample thickness and emission intensity as long as the film remained thinner than 80 nm.

The infrared emission observation is superior to the reflection measurement for obtaining a spectrum of a thin film on a rough or a curved solid surface.[20] Nagasawa and Ishitani compared the reflection spectra of poly (acrylonitrile)–polystyrene copolymer films on 2 µm-diameter copper wires with the emission spectra of the same sample.[35] The emission spectra showed higher S/N ratios than those of the reflection spectra. Lauer and King[36] used emission spectroscopy to study the effect of the addition of chlorinated hydrocarbon on an elastohydrodynamic contact.[36] They obtained spectra of a traction fluid in a bearing ball/diamond contact. The spectra obtained showed an increase in aromatics and olefins. The molecular orientation in the traction fluid of the contact zone is discussed.

Infrared emission spectroscopy has been utilized for monitoring working catalytic processes. A few examples are discussed here. Kember and Sheppard measured at 373–573 K infrared emission from a Pd/SiO_2 catalyst under the reaction gas flow conditions.[37] A remarkable difference in temperature was found in the oxidation reaction of CO with O_2 between the measured temperature and that of metal particles, because the reaction on the metal surface proceeded in a self-heating regime. Mantell *et al.* used a pulsed nozzle beam technique in combination with an FTIR spectrometer for the time resolved study of CO oxidation on Pt and Pd over the temperature range 730–1300 K.[38] A pulsed CO flux was supplied to the hot metal surface, and the emission of the reaction product CO_2 scattered from the metal surface was measured as a function of time after the pulse. Whereas no oxygen coverage effect was observed on Pd, the depletion of oxygen from the Pt surface resulted in a remarkable change in shape of the emission band of the scattered CO_2.

Sullivan *et al.*[39] reviewed the IR emission observations performed with a gas flow cell similar to that employed in the work of Mazzarese *et al.*[30] A relatively thick oxide or zeolite catalyst layer (~20 μm thick) is placed on a stainless steel stage in the cell, and the near-normal emission is collected under the gas flow. This technique will be especially useful for the simultaneous observation of the change in species on a catalyst and that in the structure of the solid catalyst during the reaction at elevated temperatures.

References

1. J. B. Bates, in: *Fourier Transform Infrared Spectrosopy, Applications to Chemical Systems,* Vol. 1 (J. R. Ferraro and L. J. Basile, eds.), Academic, New York (1978), p. 99.
2. N. Sheppard, in: *Analytical Application of FT-IR to Molecular and Biological Systems* (J. R. Durig, ed.), Reidel, Dordrecht, Holland (1980), p. 125.
3. R. P. Eischens and W. A. Pliskin, *Advan. Catal.* **10,** 51 (1958).
4. M. J. D. Low and H. Inoue, *Anal. Chem.* **36,** 2397 (1964).
5. M. J. D. Low and I. Coleman, *Spectrochim. Acta* **22,** 369 (1966).
6. P. R. Griffiths, *Appl. Spectrosc.* **26,** 73 (1972).
7. G. Fabbri and P. Baraldi, *Appl. Spectrosc.* **26,** 593 (1972).

8. J. F. Blanke, S. E. Vincent, and J. Overend, *Spectrochim. Acta* **32A**, 163 (1976).
9. J. F. Blanke and J. Overend, *Spectrochim. Acta* **32A**, 1383 (1976).
10. D. Kember, D. H. Chenery, N. Sheppard, and J. Fell, *Spectrochim. Acta* **35A**, 455 (1979).
11. R. G. Greenler, *Surf. Sci* **69**, 647 (1977).
12. R. G. Greenler, *J. Vac. Sci. Technol.* **12**, 1410 (1975).
13. K. Makinouchi, K. Wagatsuma, and W. Suëtaka, *J. Spectrosc. Soc. Jpn.* **29**, 23 (1980).
14. H. Momose, M. E. Thesis, Tohoku University (1988).
15. D. L. Allara, D. Teicher, and J. F. Durana, *Chem. Phys. Lett.* **84**, 20 (1981).
16. S. Chiang, R. G. Tobin, and P. L. Richards, *J. Electron Spectrosc. Relat. Phenom.* **29**, 113 (1983).
17. P. L. Richards and R. G. Tobin, in: *Vibrational Spectroscopy of Molecules on Surfaces* (J. T. Yates, Jr. and T. E. Madey, eds.). Plenum, New York (1987), p. 417.
18. K. Wagatsuma, K. Monma, and W. Suëtaka, *Appl. Surf. Sci.* **7**, 281 (1981).
19. K. Ohta and H. Ishida, *Appl. Opt.* **29**, 2466 (1990).
20. D. Kember and N. Sheppard, *Appl. Spectrosc.* **29**, 496 (1975).
21. T. Wadayama, W. Shiraishi, A. Hatta, and W. Suëtaka, *Appl. Surf. Sci.* **44**, 43 (1990).
22. K. Tochigi, H. Momose, Y. Misawa, and T. Suzuki, *Appl. Spectrosc.* **46**, 156 (1992).
23. R. G. Tobin, S. Chiang, P. A. Thiel, and P. L. Richards, *Surf. Sci.* **140**, 393 (1984).
24. S. Chiang, R. G. Tobin, P. L. Richards, and P. A. Thiel, *Phys. Rev. Lett.* **52**, 648 (1984).
25. R. G. Tobin and P. L. Richards, *Surf. Sci.* **179**, 387 (1987).
26. B. E. Hayden and A. M. Bradshaw, *Surf. Sci.* **125**, 757 (1983).
27. R. G. Tobin, R. B. Phelps, and P. L. Richards, *Surf. Sci.* **183**, 427 (1987).
28. K. Wagatsuma and W. Suëtaka, *J. Spectrosc. Soc. Jpn.* **30**, 258 (1981).
29. T. Wadayama, H. Mizuseki, and A. Hatta, *Vibrational Spectrosc.* **2**, 239 (1991).
30. D. Mazzarese, K. A. Jones, and W. C. Conner, *J. Electron. Mater.* **21**, 329 (1992).
31. J. L. Lauer and P. Vogel, *Appl. Surf. Sci.* **18**, 182 (1984).
32. M. J. D. Low, K. H. Brown, and H. Inoue, *J. Colloid Interface Sci.* **24**, 252 (1967).
33. L. M. Gratton, S. Paglia, F. Scattaglia, and M. Cavallini, *Appl. Spectrosc.* **32**, 310 (1978).
34. T. Matsui, K. Tani, S. Ohashi, and T. Tanaka, *J. Spectrosc Soc. Jpn.* **31**, 360 (1982).
35. Y. Nagasawa and A. Ishitani, *Appl. Spectrosc.* **38**, 168 (1984).
36. J. L. Lauer and V. M. King, ASLE Preprint 80-AM-4D-2 (1980).
37. D. R. Kember and N. Sheppard, *J. Chem. Soc., Faraday Trans. 2*, **77**, 1321 (1981).
38. D. A. Mantell, S. R. Ryali, and G. L. Haller, *Chem. Phys. Lett.* **102**, 37 (1983).
39. D. H. Sullivan, W. C. Conner, and M. P. Harold, *Appl. Spectrosc.* **46**, 811 (1992).

Surface Raman Spectroscopy

1. Introduction

Raman scattering, an inelastic scattering of photons by chemical entities, is widely utilized for the investigation of the structure of molecules, crystals, and polymer materials.[1–5] The cross section of normal Raman scattering, however, is low for observing surface species, and most surface Raman experiments were confined to porous or finely divided inorganic materials until recently.[6,7] Before the discovery of surface enhanced Raman scattering (SERS), resonance Raman scattering[8,9] was employed for the most part in the observation of bulk solid surfaces.[10]

While the discovery of SERS has greatly stimulated the investigation of solid surfaces with the enhanced Raman technique, the use of normal and unenhanced Raman scattering for the observation of solid surfaces has steadily increased as a result of the advance in technology of the detection of faint signals of Raman scattering as well as in the understanding of the optimum conditions for observation. The investigation of solid surfaces with unenhanced Raman spectroscopy has been reviewed recently by Campion.[11,12] In this chapter, mainly normal Raman spectroscopy is treated and SERS will be dealt with in the next chapter.

Raman scattering provides us with spectra complementary to the infrared absorption, because the mutual exclusion rule is generally applicable to molecules having a center of symmetry, i.e., infrared active vibrations are Raman inactive and vice versa. Raman spectroscopy has several advantages for the investigation of solid surfaces. It can be used for the observation of materials in vacuum as well as in ambient conditions, as is the case with IR spectroscopy, enabling us to observe *in situ* the behavior of species on catalysts and electrodes. It is superior to IR absorption measurements for the *in situ* observation of electrode surfaces, because the Raman scattering of the solvent brings about much less serious problems than the overwhelming IR absorption of the solvent. Raman spectroscopy provides information over a wider wavenumber region than the IR observation in a single measurement. The spatial resolution of Raman spectroscopy is much better than that of IR measurement because of the shorter wavelength of the probing light. Furthermore, the totally symmetric vibrations can be identified from the measurement of the depolarization ratios of Raman lines.

There also exist several drawbacks in Raman spectroscopy. Besides the intrinsic low sensitivity, emission from hot samples and fluorescence are often very troublesome in the observation of Raman spectra. The low sensitivity of normal Raman spectroscopy may be overcome by the use of resonance (or preresonance) Raman scattering, when the sample is colored. However, the continuous irradiation of the sample with an intense laser beam of a wavelength lying inside the absorption band of the sample results in the decomposition of the sample compound. Although spinning the sample cell can help minimize the decomposition, this scheme is not easily practicable for the measurement of functioning electrode and catalyst surfaces.

Irradiation by intense visible or near-ultraviolet laser beams may damage the sample compounds, even if they are uncolored. The reduction in intensity of the exciting light lessens the possibility of damage but inevitably impairs the quality of the measured spectra. The decrease in intensity of the exciting light can be compensated for by the increase in number of sample molecules: The scanning of the laser beam[13] or the spreading of the exciting beam[14] over the sample surface may be effective for obtaining good spectra when the sample has a large, flat surface.

The multichannel detection of Raman signals[3,6] is another potential alternative for improving the S/N ratio of the spectrum. Campion has shown that, when used in combination with a cooled charge-coupled device detector (CCD), a single spectrograph equipped with Bragg diffraction notch filters or a triple spectrograph may be effectively employed for surface Raman measurements.[11,12] Fourier transform-Raman spectroscopy possesses the multiplex advantage as well.[15] However, a near-infrared laser is used at present as the exciting source of FT-Raman spectrometers. This system is effective for removing fluorescence, but gives sensitivity insufficient for the observation of unenhanced Raman scattering from surface species, because the intensity of Raman lines changes approximately in proportion to the fourth power of the frequency of the exciting light and the near-infrared photon detectors are rather low in sensitivity.

A potential modulation method and a difference Raman technique have been applied to the *in situ* observation of species generated electrochemically.

Waveguide and total reflection Raman spectroscopy have been utilized successfully for the investigation of thin films, surface layers, and interfaces. These techniques are dealt with in this chapter.

2. Optimum Conditions for the Measurement of Surface Raman Spectra

2.1. External Reflection Method

Since the signal from normal Raman scattering is rather weak, we must know the experimental conditions under which the maximum scattering intensity is obtained. The intensity, I_{mn}, per unit solid angle of Raman scattering stemming from the transition from an initial state, m, to final one, n, is given by

$$I_{mn} = \text{const} \, (v_0 - v_{mn})^4 I_0 \sum_{\rho\sigma} |(\alpha_{\rho\sigma})_{mn}|^2 \qquad (1)$$

where I_0 and v_0 are the intensity and the frequency of the exciting light, and v_{mn} is the frequency corresponding to the energy difference between the final and initial states. $\alpha_{\rho\sigma}$ is the $\rho\sigma$ component of the Raman scattering tensor and can be written in the form

$$(\alpha_{\rho\sigma})_{mn} = \sum_{r} \left(\frac{(M_\sigma)_{mr}(M_\rho)_{rn}}{E_r - E_m - hv_0} + \frac{(M_\rho)_{mr}(M_\sigma)_{rn}}{E_r - E_n + hv_0} \right) \qquad (2)$$

where E_r, E_m and E_n are the energy of the intermediate (r), initial, and final states, respectively. $(M_\sigma)_{mr}$ is the σ component of the transition moment associated with the m and r states, and σ and ρ denote x, y, and z components.

From Eq. (1) we see that the intensity of Raman scattering can be enhanced by increasing $(v_0 - v_{mn})$, I_0, and the component of the Raman scattering tensor. The frequency of Raman shift, v_{mn}, is small in comparison with the frequency, v_0, of the incident exciting light. The highly monochromatic radiations in the visible or near-ultraviolet regions from gas lasers therefore, are used as the source for the excitation of Raman scattering. The gas lasers give a self-collimated beam having high intensity, which in turn brings about an increase in the sensitivity of the Raman spectroscopy. In the surface Raman measurement, however, I_0 is not the intensity of the incident laser beam but rather the strength of the surface electric field, which may be enhanced in the external and internal total reflection measurement.

The last factor increases in the resonance Raman scattering. The denominator of the first term in the bracket of Eq. (2) can be written as

$$E_r - E_m - h\nu_0 = h(\nu_{rm} - \nu_0) \tag{3}$$

It follows, then, that when the frequency of the incident beam approaches ν_{rm}, which is the frequency corresponding to an electronic transition from the electronic ground state to the excited state r, $\alpha_{\rho\sigma}$ takes a large value, provided the transition moments, $(M_\sigma)_{mr}$ and $(M_\rho)_{rn}$, have nonzero values: in other words, the transition between the electronic ground state and the excited state, r, is allowed. This is the origin of the increase in scattering intensity of resonance Raman spectra. This mechanism is operative also in some "chemically" enhanced SERS.

In the external reflection measurement of Raman spectra, there exist additional factors which determine the observed intensity of Raman lines. In a simple classical picture, Raman scattering can be divided into two process:[16] (1) the transition induced by the irradiation of the exciting light from the vibrational state, m, in the electronic ground state to an excited (virtual) state, v, and (2) the transition from the excited state to the vibrational state, n, in the electronic ground state accompanied by the spontaneous emission of Raman shifted light, as shown in Fig. 1.

In Chapter 2, we saw that the electric field strength on solid surfaces changes drastically depending upon the incident angle and the state of polariza-

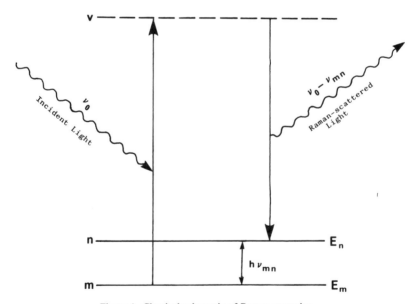

Figure 1. Classical schematic of Raman scattering.

tion of the probing IR beam as well as on the optical constants of the solid. We learned in Chapter 4 that the intensity and the polarization state of infrared emission from species on solid substrates varies remarkably with changes in the angle of observation and the optical constants of the substrate. The solid substrate plays a similar role in the Raman scattering from species on its surface. In consequence, when a laser beam is incident at the angle corresponding to maximum electric field generated on the solid surface and the scattered light is collected at the angle corresponding to the maximum intensity, the S/N ratio of Raman spectra of species on the solid surfaces can be improved; on highly reflective metal surfaces, the S/N ratio may be enhanced by nearly one order of magnitude.

The incident laser beam generates an oscillating electric field of a standing wave on the metal surface as a result of the combination with the reflected light. Greenler and Slager computed the angular dependence of the strength of the electric field generated on metal surfaces upon incidence of 488 nm laser light.[16] Mullins and Campion calculated the generated electric field strength upon incidence of p- and s-polarized light on a flat silver surface.[17] Figure 2 shows the results obtained by Greenler and Slager.[16] This figure shows only the results for p-polarized light. It is clear in this figure that the Raman

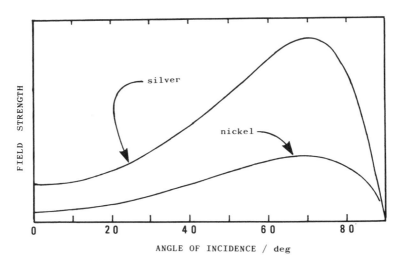

Figure 2. Calculated strength of electric field on metal surfaces generated upon incidence of p-polarized 488 nm laser light [Reprinted from *Spectrochim. Acta* **29A**, R. G. Greenler and T. L. Slager, Method for obtaining the Raman spectrum of a thin film on a metal surface, p. 194, © 1973 with kind permission from Pergamon Press Ltd, Headington Hill Hall, Oxford, OX3 0BW, UK].

excitation, assumed to be proportional to the standing wave intensity, reaches a maximum as the laser beam is incident at about 70° on the silver surface.

The angle dependence calculated by Mullins and Campion is in approximate agreement with the curves shown in Fig. 2, and shows that the electric field normal to the silver surface has the maximum strength upon incidence of p-polarized light (λ = 520 nm) at an angle of 65°. They also computed the field intensity on nickel and silicon surfaces. The intensity decreases from silver to nickel and to silicon, but qualitatively speaking, the angular distribution is the same for all of these materials. Roughly speaking the incidence of p-polarized visible light at a high angle generates on metal surfaces an enhanced electric field polarized nearly normal to the surface in agreement with the results for infrared light described in Chapter 2. However, the maximum strength of the electric field on the silver surface is only about twice the strength of the incident light, significantly smaller than the enhancement of the infrared field. Furthermore, if the coordinate system shown in Fig. 3 is used, the x-component of the standing electric field and the y-component, generated by the incidence of s-polarized light, are not negligible in comparison to the z-component, in contrast to the infrared field, but have values of about 1/10 and 1/30 of the z-component at an angle of 65°, respectively.[17] In addition, the maximum field

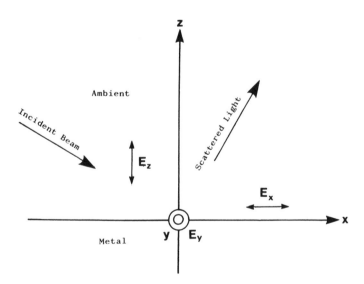

Figure 3. The coordinate system used in this chapter.

strength is obtained by the incidence of visible light at an angle lower than for infrared light.

The angular dependence of the Raman scattered light is also similar to that of the infrared emission. Figure 4 shows the dependence calculated by Greenler and Slager. The intensity distribution of the scattered light from the induced dipole oscillating normal to the metal surface is shown in this figure. The intensity of the scattered light reaches a maximum value at an angle near 60°, an angle somewhat smaller than that in the infrared emission.[16] Mullins and Campion also calculated the angular distribution of the Raman scattered light from oscillating induced dipoles oriented in x, y, and z directions shown in Fig. 3. They have taken into account the finite angular range of the collection optical system in the calculation. The calculation has shown that the maximum intensity is obtained at a collection angle of 55°. The contribution to the measured intensity comes mostly from the dipole oriented normal to the surface, but the dipoles oscillating in the x and y directions contribute appreciably to the observed intensity in contrast with the infrared emission.[17]

The optimum conditions for measuring the Raman spectrum of species on

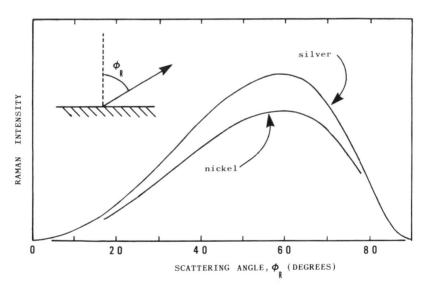

Figure 4. Calculated angular distribution of scattered light of 488 nm wavelength from an induced dipole normal to the surface [Reprinted from *Spectrochim. Acta* **29A**, R. G. Greenler and T. L. Slager, Method for obtaining the Raman spectrum of a thin film on a metal surface, p. 195, © 1973 with kind permission from Pergamon Press Ltd, Headington Hill Hall, Oxford, OX3 OBW, UK].

metal surfaces is obtained from the following discussion: a p-polarized laser beam should be incident at an angle in the range 60~70° and the scattered light should be collected at an angle around 60° from the surface normal. These conditions are optimum, however, only for the observation of species present in the vicinity closest to the metal surface, because the wavelength of the probing light generally employed in the Raman observation is about one-tenth of the infrared radiation and the strength of the standing electric field changes steeply with the distance from the surface.

Figure 5 shows Raman spectra of 7,7,8,8-tetracyanoquinodimethane (TCNQ) thin films evaporated on aluminum mirrors.[18] The s-and p-polarized laser beams of 514.5 nm in wavelength were incident on the sample at an angle of 70°, and the scattered light was detected using collection angles of 0° and 60°. Although the sample used was not the best for confirming the theoretical prediction, because TCNQ films were not isotropic and the Al mirrors were covered with native oxide, the best quality spectrum was obtained with the optimum conditions predicted theoretically for a film of 6 nm thickness, as can be seen in Fig. 5a. These optimum conditions, however, change with increasing film thickness, and the use of a collection angle of 0° gives Raman lines slightly stronger than those obtained at a collection angle of 60° in the measurement of a film of 15 nm in thickness.

The field strength in transparent films on metal surfaces is enhanced as a result of the interference of multiply reflected light. Lenac and Tomaš made a theoretical investigation of the interference effect in Raman scattering from overlayers on metals using a four-layer model.[19] Besides the effect of metal thickness, they showed the signal enhancement in a sandwich-type configuration. The application of an optically dense material (prism) to the transparent dielectric layer on the metal further enhances the field in the dielectric, and strong enhancement of Raman signals may be obtained by embedding the sample molecules in the dielectric at the optimum distance from the metal surface in a prism/dielectric/metal configuration. The interference enhancement of the field strength in the ATR configuration occurs in the infrared region as well and is treated in Section 2.2 of Chapter 3. It should be noted that Raman scattering from the prism may cause a serious problem, which will be treated in the next section.

The surface selection rule for the absorption and emission of infrared radiation is very useful for the determination of the orientation of species on metal surfaces. It would be convenient in the investigation of surface species to have this advantage also for the Raman scattering. However, the selection rule for the Raman scattering is much more complicated than for infrared spectroscopy, even if only unenhanced Raman spectroscopy is concerned, because the selection rule is associated with the dipole moment change in the

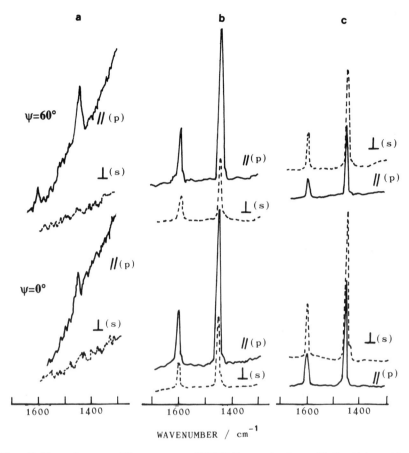

Figure 5. Change in measured Raman spectra of TCNQ films on aluminum with film thickness (*d*) and angle of observation (ψ); *p*- and *s*-polarized 514.5 nm laser beams were incident at 70° (a) d = 6 nm, (b) d = 15 nm, (c) d = 50 nm [Reprinted from *Spectrochim. Acta* **36A**, M. Osawa, W. Kusakari, and W. Suëtaka, Raman spectra and molecular orientation of 7, 7′, 8, 8′-tetracyanoqui-nodimethane in thin films evaporated on aluminum, p. 390, © 1980 with kind permission from Pergamon Press Ltd, Headington Hill Hall, Oxford, OX3 OBW, UK].

infrared absorption and emission, while in Raman spectroscopy it depends upon the polarizability change. Whereas the surface field polarized normal to the metal surface induces infrared absorptions of the dynamic dipoles oriented normal to the surface, the surface field polarized likewise induces dipoles not necessarily oriented perpendicular to the surface in Raman scattering.

Moreover, the presence of the component parallel to the metal surface in the standing electric field and the appreciable contributions from induced dipoles oriented parallel to the surface to the measured intensity of the scattered light make difficult the formulation of any straightforward selection rule for surface Raman spectroscopy. In spite of such a difficulty, a loose surface selection rule, called the propensity rule, is applicable to unenhanced Raman scattering:[11,12,20,21] the vibrational modes transforming as α_{zz}, which is a component of the polarizability tensor based on the surface-fixed axes with z surface-normal, are expected to appear strongly, while modes transforming as xz and yz are observable and xy, xx, and yy modes should be very weak.[11]

It would be worthwhile to mention here the symmetry-lowering of species upon adsorption on a surface of single crystals and the resulting appearance of modes that are otherwise Raman inactive.[11] Figure 6 shows Raman spectra of C_6D_6 physisorbed on Ag[111] and Ag[110] surfaces. Benzene-d_6 adsorbed on

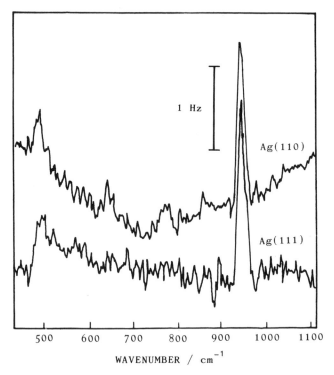

Figure 6. Raman spectra of C_6D_6 physisorbed on Ag[111] and Ag[110] at 90 K and submonolayer coverage (adapted from Ref. 11).

Ag[111] shows two bands at 945 (A_{1g}) and 495 (A_{2u}) cm^{-1}. The latter band is Raman inactive for free benzene, and its appearance in the spectrum shown in this figure is due to the symmetry-lowering from D_{6h} to C_{6v} resulting from the benzene adsorption with its plane parallel to the surface and the resultant loss of a symmetry plane. The spectrum of benzene adsorbed on Ag[110] shows two additional features at 780 and 650 cm^{-1}, assignable to vibrations belonging to E_{2u} and E_{1g} species, respectively. The appearance of the former feature indicates a symmetry reduction to C_{2v}. These kinds of changes of spectrum that depend upon the crystal face can be used for the investigation of the adsorption site.[11]

2.2. Total Reflection Raman Spectroscopy

The external reflection method is not effective for the observation of Raman spectra of surface species on transparent substrates, because the electric field on the substrate generated by the incidence of a laser beam is significantly weaker than on metal surfaces. An alternative for obtaining good Raman spectra of surface species is the total internal reflection method, which has been utilized for the measurement of Raman spectra of the surface layer of liquids,[22-25] polymer films,[26-28] and liquid/liquid interfaces.[29] Since the penetration depth in the total reflection Raman measurement is much smaller than in the infrared ATR measurement, the Raman measurement is adequate for the investigation of thin surface layers.

The sensitivity of Raman spectra increases with I_0 in Eq (1), which is the strength of the surface field in the surface Raman excitation. In the total internal reflection arrangement, the strong surface electric field of the evanescent wave is obtained by the incidence of light at the critical angle, as described in Chapter 3. The s-polarized light of unit strength generates surface field, polarized parallel to the prism surface, of a maximum value of 4 upon incidence at the critical angle, while the p-polarized light generates a maximum evanescent field of $4 \times (n_1/n_2)^2$, polarized normal to the prism surface, where n_1 and n_2 are the refractive indices of the prism and the rarer medium, respectively.

The angular distribution of the intensity of Raman scattered light in the total reflection measurement should be similar to that of the emission from species located near the prism surface. The angular distribution of the light emitted incoherently from dipoles located in close vicinity to a plane dielectric in investigated theoretically and a sharp angular distribution was ca ') The fluorescence emitted from a thin film evaporated on a pri howed a sharp angular distribution.[33] These results imply that the Raman scattered light should show a sharp angular distribution in nal reflection measurement.

Mattei *et al.* formulated a semiclassical theory for the efficiency of Raman scattering in the total internal reflection arrangement, and deduced that the strongest Raman signal from the isotropic solid in contact with the face of an ATR prism is obtained when both the incident angle of the exciting beam and the collection angle of the scattered light are equal to the critical angle.[34] They confirmed the prediction in the measurement of Raman spectra from the sapphire prism–$NaBrO_3$ crystal sample system. Ohsawa *et al.* investigated the angular distribution of Raman scattered light from a thin film of oriented molecules on a prism base.[35] The observed intensity of the Raman signal was a sharp function of collection angle and the maximum intensity was obtained independent of the molecular orientation when the scattered light was collected near the critical angle through the prism. The optimum conditions for the total reflection Raman measurement of isotropic thin films can be drawn from these studies: a *p*-polarized laser beam should be incident at the critical angle and the Raman signal scattered into the prism should be collected at the critical angle. When the sample molecules are oriented, the measurement under these conditions is effective for the observation of Raman bands transforming as *zz*, *xz*, and *yz*, and a very simple spectrum may be obtained, as is often the case with the reflection infrared measurement of oriented species on metal substrates. The use of *s*-polarized light is adequate for the observation of Raman signals of the modes transforming as *yy* and *xy*, and for the observation of the modes transforming as *xx*, the incidence of *p*-polarized light at an angle somewhat higher than the critical angle is preferable so as to obtain the highest exciting field in the *x*-direction.

The total internal reflection method, thus, can be used for the observation of Raman spectra of thin films on transparent substrates. When the substrate has a high refractive index and shows only weak Raman lines in the wavenumber region of interest, it can be used as the internal reflection element; otherwise a hemicylinder prism may be put into contact with the surface of the film or the backsurface of the substrate, which should be parallel-sided and thin. A small quantity of liquid may be applied to the interface between the substrate and the prism. Care must be taken in selecting the material of the prism, because strong signals of Raman scattering of the prism often give rise to serious problems in the detection of faint signals of surface species. Sapphire ($n = 1.77$) is often chosen as the prism material,[28,34] because its strong Raman lines lie below 760 cm^{-1}. Optical glasses having high refractive indices, such as LaSF15 ($n = 1.88$), are used for the prism as well.[35]

Another alternative to obtaining Raman spectra of thin films is the waveguide method.[36] The intensity of Raman lines increases with an increase in the scattering volume of the sample material. High quality Raman spectra of transparent films would be obtained if the exciting laser beam could be introduced into the film and could propagate inside the film through multiple internal reflection. However, when the film thickness is smaller than the critical value, the exciting beam does not propagate within the film. In order to overcome this limitation, Levy *et al.* fabricated an optical waveguide combining the sample film with a transparent thin solid substrate for the laser beam to propagate through the sample and observed Raman spectra of polymeric films.[37] Unfortunately, the S/N ratio of the obtained spectra was rather poor because of the superimposition of signals from the substrate.

Rabolt *et al.* obtained Raman spectra with good S/N ratios of 1 μm polymer films with the waveguide method.[27,38] They employed the asymmetric slab waveguide shown in Fig. 7.[38] The waveguide consists of a sample polymer film, a thin layer of Corning glass ($n = 1.56$) sputtered onto a Pyrex ($n =$

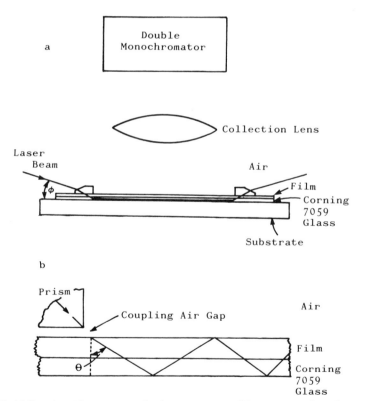

Figure 7. (a) Experimental arrangement for the measurement of Raman spectra from the waveguide; (b) schematic of propagating wave in combined polymer-Corning glass layer [Reprinted from J. F. Rabolt, R. Santo, and J. D. Swalen, *Appl. Spectrosc.* **34**, 517 (1980)].

1.48) substrate, and a superstrate (air). The organic film is deposited directly on the Corning glass layer. The laser beam is coupled by a prism of high refractive index optical glass (LaSF5) into the combined layer of the waveguide. The Raman scattered light induced by the guided wave is collected with the lens shown in the figure.

The optical field intensity within the waveguide can be changed by changing the coupling angle. Fig. 8a shows the optical field intensity across the poly(vinyl alcohol) (PVA)-poly(methyl methacrylate) (PMMA) bilayer polymer waveguide formed by the incidence of an s-polarized beam.[39] The mode number, m, is related to the discrete coupling angles and gives the number of nodes in the field distribution. The field intensity in each polymer layer changes

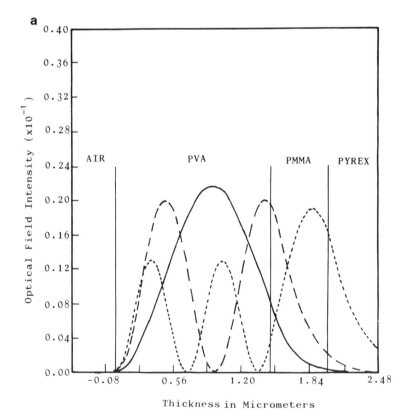

Figure 8. (a) Distribution of field intensity across the PMMA/PVA layer on Pyrex for three modes; ——, $m = 0$; – – –, $m = 1$; - - - -, $m = 2$; $n_{PVA} = 1.5224$, $n_{PMMA} = 1.4945$, $n_{PX} = 1.4706$.

with the mode number: the $m = 0$ mode gives a field nearly confined in the PVA layer of the polymer laminate, and the strong fields in the PMMA layer and at the PVA/PMMA interface are obtained in the $m = 2$ and $m = 1$ modes, respectively. Raman spectra obtained by changing the coupling angle and consequently the mode number are shown in Fig. 8b. The figure shows clearly that the spectrum of the polymer laminate changes corresponding to the field distribution. This figure demonstrates the feasibility of waveguide Raman spectroscopy for the investigation of thin polymer films, polymer/polymer and polymer/glass interfaces, the interlayer transport between two films, and so on.

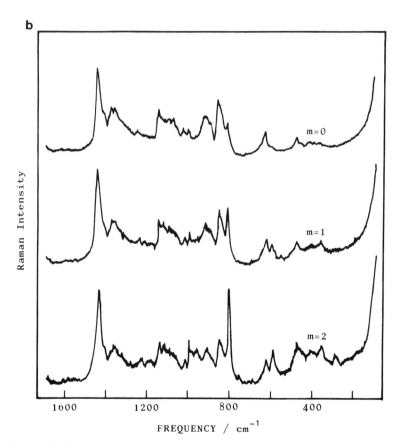

Figure 8. (b) Raman spectra of the PMMA/PVA layer corresponding to the field distribution shown in (a) [Reprinted with permission from J. F. Rabolt, N. E. Schlotter, and J. D. Swalen, *J. Phys. Chem.* **85**, 4141 (1981). © 1981 American Chemical Society].

The irradiation of a thin film with a visible laser beam often gives rise to sample degradation or fluorescence that often buries weak Raman signal. To eliminate this difficulty, an FT-Raman technique, where near-infrared lasers are used for the excitation, has been introduced into the waveguide Raman measurement.[40] A fiber optic bundle was used in this work instead of a collection lens for collecting the scattered light, and an improvement in the S/N ratio was obtained. However, the use of the near-infrared light as the excitation source gives rise to several serious problems. The efficiency in inducing Raman scattering decreases remarkably, and the critical thickness of the waveguide increases. The refractive index of polymeric materials decreases significantly with increasing wavelength, and the distribution of the optical field across the film becomes broad. Zimba and Rabolt discussed the choice of the exciting source and the resulting effect on the waveguide Raman measurement.[41]

It should be added that some reports have appeared in the literature on the combination of wave guide Raman spectroscopy with coherent anti-Stokes Raman scattering (CARS).[42,43]

3. Applications

3.1. Thin Films on Solid Surfaces and Surface Layers

Both the external reflection and internal total reflection methods are employed for the observation of Raman spectra of adsorbed species and thin films on solid substrates. The surface and interface layers are mainly observed with the total reflection and waveguide methods. The backscattering Raman measurement is often used for the investigation of surface layers of deeply colored solid materials.

Until the mid-1970s, high area powders were used as the substrates in the observation of normal Raman spectra of adsorbed species.[44–47] After Greenler and Slager determined the optimum conditions for reflection measurements,[16] the external reflection method was applied to the investigation of thin films and adsorbed species on bulk metal surfaces.

Howe *et al.* employed these optimum conditions in their Raman measurement of polystyrene thin films on a silver mirror. A Raman band was barely discernible in the spectrum obtained from a film of 5 nm in thickness.[48] Stencel and Bradley were able to measure unenhanced Raman spectra of CO adsorbed at reduced pressures on low-index faces of a Ni single crystal.[49–51] In subsequent work, Bradley *et al.* were successful in obtaining Raman spectra of CO adsorbed at 98 K on a well defined Ni[111] surface through the aid of a data processing facility, and the signals of C–O and Ni–C stretching vibrations for both linear and bridge-bonded CO were located.[52]

Campion *et al.* investigated the feasibility of unenhanced Raman spectroscopy for the investigation of adsorption and chemical reactions on metal surfaces. They chose nitrobenzene as the adsorbed compound because of its large cross section of Raman scattering and observed Raman spectra from sample molecules adsorbed on Ni[111] in ultrahigh vacuum at 100 K. A cooled, intensified vidicon was used for the multichannel detection of the dispersed radiation, and the signals were accumulated for 500–1000 s. The spectrum of nitrobenzene at submonolayer coverage was different from that of the condensed film, as shown in Fig. 9, indicating the dissociative adsorption of nitrobenzene to form nitrosobenzene adsorbed on the Ni surface.[11,53]

Defects on metal surfaces may play an important role in the "chemical" mechanism of surface enhanced Raman scattering. Campion and Mullins observed Raman spectra of pyridine adsorbed on stepped and kink-stepped silver surfaces in ultrahigh vacuum.[54] No enhancement, however, was detected in Raman scattering from pyridine molecules physisorbed on stepped and kink-stepped surfaces of Ag[521] and Ag[987] or the same molecules chemisorbed on Ag[540], showing that the chemisorption at the surface defects alone is not sufficient to produce the "chemical" enhancement. Pyridine molecules chemisorbed on Ni[111] and Ni[100] surfaces exhibited interesting differences in their Raman spectra.[55] Whereas Raman signals were obtained from pyridine on the Ni[100] surface at a coverage of 0.5 ML, no Raman line was detectable in the spectra of the Ni[111] surface. This difference has been attributed to a phase transition and the resultant change in molecular orientation, which increases the Raman intensity and which does not occur on the Ni[111] surface at a coverage of 0.5 ML.

The feasibility of unenhanced Raman spectroscopy for the observation of species on semiconductor surfaces was investigated by Shannon and Campion,[56] because important surface processing reactions of semiconductors are visible laser-mediated and the surface fields formed on Si and, most probably, on the other semiconductor surfaces, upon incidence of a laser beam have sizable strength.[11] They observed spectra of acetonitrile and methanol on the Si[100]2 × 1 surface at 100 K in ultrahigh vacuum. Spectra of thin films of these compounds in the wavenumber region where the substrate did not scatter strong signals could be obtained, and the orientation of these molecules in the films could be discussed qualitatively. On the other hand, a Raman band was discernible in the spectra of the acetonitrile monolayer on Si, but no band was located in the spectrum of the methanol monolayer.

The backscattering method is employed for the observation of Raman

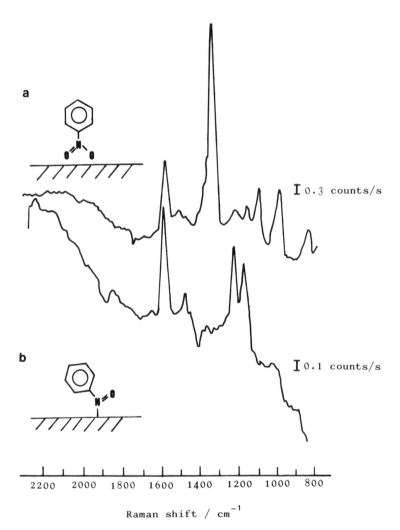

Figure 9. Raman spectra of nitrobenzene adsorbed at (a) multilayer and (b) submonolayer coverages on Ni[111] at 100 K. 514.5 nm laser light was used for the excitation at a power of 150 mW (adapted from Ref. 11).

spectra of the surface layers of semiconductors. A laser beam is incident at a high angle and the scattered light is collected in a cone normal to the surface. In this arrangement, the spectrum of a surface layer of the optical penetration depth (usually less than 1 μm for semiconductors) can be obtained. The presence of a microcrystalline Te layer was found on the surface of CdTe etched in dilute mineral acids by the Raman measurement.[57] Raman spectra of surface layers of PbSnTe freshly cleaved in air or etched electrolytically showed the presence of a very thin oxide (TeO_2) film, which changed with time to a layer of mixed oxide of Pb and Te. On the other hand, the surfaces etched in HNO_3 and HCl gave spectra of thin layers of Te.[58]

The Raman band of the surface layer on amorphous silicon crystallized by laser annealing appeared at lower frequency than the crystal silicon. This difference has been attributed to strain in the surface layer.[59] Stolz and Abstreiter noticed a difference in intensity of a forbidden LO Raman band of highly doped GaAs between the samples cleaved in ultrahigh vacuum and those broken in air. The appearance of the forbidden mode and the difference in intensity may be attributed to the high electric field in the depletion layer at the surface and its difference between two samples.[60] The Raman spectra of GaAs covered by thermally grown oxide revealed the presence of elemental arsenic at the interface. The Raman spectra of anodically oxidized GaAs showed at room temperature the LO and TO modes of the GaAs substrate, but the sample exhibited strong LO and TO bands of crystalline arsenic and broad features due to amorphous arsenic after heating to 723 K or above. The presence of elemental arsenic at the interface was ascribed to the interfacial reaction between the surface oxide and the substrate.[61]

Raman spectra of the high T_c superconductor $YBa_2Cu_3O_7$ and the semiconductor $YBa_2Cu_3O_6$ were observed with the backscattering method. The feature assignable to the symmetric stretching of the terminal Cu–O bonds in the CuO_3 units of $YBa_2Cu_3O_7$ was shifted to a much lower frequency in the nonsuperconducting compound $YBa_2Cu_3O_6$.[62]

Raman spectra of surface layers of solid materials can also be obtained with the total reflection and waveguide methods. Mattei *et al.* observed a surface phonon polariton mode in addition to the bulk TO and LO modes of $NaClO_3$ using a sapphire prism-$NaClO_3$ system.[63] Raman spectra of thin polymer coatings can easily be measured with the total internal reflection method. The spectrum of a 1.1-μm polystyrene coating on a polyethylene film (30 μm thick) was obtained without signals from the underlying film.[28] The internal reflection method was utilized for the measurement of resonance Raman spectra of species present at the water/CCl_4 interface.[29] Tian *et al.* observed resonance Raman spectra of azobenzene-containing amphiphile monolayers adsorbed at the acidic aqueous solution/CCl$_4$ interface, and they discussed the molecular aggregation as well as the orientation of protonated species in the adsorbed monolayer.[64]

The waveguide Raman method was applied to the observation of sub-micron-thick polymer layers on glass substrates.[38,65] Raman spectra of a Lang-muir–Blodgett film of 13 monolayers of cadmium arachidate on a Corning 7059 glass were also measured as a function of temperature by the same method.[66]

Polyimide forms pinhole-free and heat resistant films and is widely used in the microelectronics industry However, understanding is still lacking of the chemical bonding within the film and at the interface with the substrate. Perry and Campion investigated the reaction of pyromellitic dianhydride (PMDA) and oxydianiline (ODA) to form polyimide on Ag[110] surfaces in vacuum with the external reflection method.[67] Oxydianiline was physisorbed to form a condensed multilayer on the silver surface at 140 K, but in contrast, PMDA was adsorbed dissociatively. Upon heating to 473 K, the coadsorbed film of these two compounds converted into the polyimide film exhibiting character-istic features in the carbonyl stretching region of the Raman spectrum. Perry and Campion also studied the nature of adhesion of the polyimide film to various substrates.[68] For investigating the interaction of PMDA and ODA with the substrates, Raman spectra were observed in the course of solvent-free vapor deposition on Ag[110], Si[100], and silicon oxide surfaces. The obtained spec-tra imply that, whereas PMDA interacts strongly with the substrates and forms various carboxylate species on the surface, ODA is adsorbed weakly through the phenyl ring on the Si[100] surface and physisorbs on silver and silicon oxide. They have also found the formation of bidentate bridging and chelating carboxylate species on Si[100] in the dissociative adsorption of PMDA. The oxidation of this surface depresses the bidentate chelate formation through blocking of the active site, but the formation of bridging carboxylate continues even on thick oxide layers.[69]

3.2. Electrode Surface and Corrosion Products

In situ observation of vibrational spectra of electrode surfaces provides us with information indispensable for the understanding of the events proceeding on the electrode surfaces. Raman spectroscopy seems to be quite feasible for the *in situ* observation, because the most frequently used solvent, water, ab-sorbs very strongly the infrared light but is transparent in the visible region and the Raman scattering of water is weak. However, the sensitivity of unenhanced Raman spectroscopy is not sufficient for the *in situ* observation of trace species on the electrode. This requires not only the use of species having large Raman cross section as the sample material but also an increase in the Raman scattering volume in addition to the use of a sophisticated detection system.

The first observation of working electrode surfaces with unenhanced Raman spectroscopy was done by Fleischmann et al.[70] using an Hg/Pt electrode in aqueous solutions containing HCl, HBr, and NaOH, and Raman bands of Hg_2Cl_2, Hg_2Br_2, and HgO were located. The species detected in this work were present as thin layers on the electrode and had particularly large cross sections of Raman scattering. To extend the Raman measurement to the other species adsorbed on electrodes, they used electrodes with a high surface area in order to increase the scattering volume and obtained strong Raman signals from pyridine adsorbed on an electrochemically roughened silver electrode.[71] However, the observed features did not stem from the unenhanced Raman scattering of pyridine molecules; this was the first observation of a surface enhanced Raman spectrum. For some time the unenhanced Raman observation was still confined to thin surface films of strong Raman scatterers. Reid et al.[72] investigated the corrosion of a lead electrode in aqueous chloride media with normal Raman spectroscopy and found the presence of the $(PbOH)_nCl_n$ phase that was missing in the potential-pH diagram (Pourbaix diagram[73]).

Loo and Lee studied electrochemical processes on gold and platinum electrodes in halide solutions by means of Raman spectroscopy and detected gold-halide complexes (AuX_2^- and AuX_4^-) on the gold electrode and halogens (Cl_2, Br_2, and I_2), as well as trihalide ions (Br_3^- and I_3^-) on Pt electrodes under anodic polarizations.[74,75] The unenhanced Raman spectrum of 4-cyanopyridine adsorbed on a rhodium electrode was observed by Shannon and Campion in an aqueous solution.[76] Whereas the p-polarized laser light gave a spectrum of the sample molecules, no Raman band was located with incident s-polarized light. This implies that the measured spectrum comes from the adsorbed species. The authors have deduced that 4-cyanopyridine is bound to the electrode through the ring nitrogen and oriented with its ring perpendicular to the surface. The maximum intensity of the Raman signals occurred at potentials in the double-layer region, showing completion of saturation coverage at the potentials.

In situ Raman observation of electrode surfaces can be fruitful for the fundamental study of metallic corrosion and passivity, and new techniques have been developed for the investigation of corrosion and related phenomena. Melendres et al.[77] extended the *in situ* observation of Raman spectra of electrode surfaces into high-temperature/high-pressure aqueous environments. They studied the corrosion and passivation of lead electrodes in dilute sulfate solution at temperatures up to ~563 K and at a pressure of 10^7 Pa. They found in this measurement the formation of the passive film of $PbSO_4$ at a potential

−0.10 V (versus Ag/Ag$_2$SO$_4$). Melendres applied another new technique of difference Raman spectroscopy to the *in situ* observation of electrode surfaces.[78] In this method, two electrodes are placed in an electrochemical cell and used as the sample and reference electrodes. The reference electrode is kept free of surface film by the application of an appropriate potential. Each electrode is illuminated by half of the incident laser beam. The scattered light from the electrodes is focused with a collection lens onto the slit of a triple monochromator, and the dispersed light is detected with a photomultiplier tube that includes amplifying microchannel plates and is capable of two-dimensional imaging.[79,80] Half of the two-dimensional image comes from the light scattered from the sample electrode and the other half from that of the reference. The images are plotted out separately and subsequently subtracted to obtain the spectrum of the surface film. The spectrum of a phosphate film on an iron electrode in an aqueous NaH$_2$PO$_4$ solution was obtained with this technique.[78] The difference Raman spectrum can also be measured by the potential modulation method, which will be described below.

For the *in situ* observation of electrochemical reactions in which colored species are produced, a resonance Raman measurement is widely used. Jeanmaire *et al.* employed the resonance Raman measurement for investigating electrogenerated intermediates and products in bulk solution and in the diffusion layer.[81] They observed *in situ* the formation of tetracyanoethylene (TCNE) anion radical during the electroreduction of TCNE in acetonitrile. The same group also measured the resonance Raman spectrum of a decay product in the reaction of electrogenerated TCNQ^{2-} with oxygen and found that the decay product was α, α-dicyano-*p*-toluoylcyanide ion.[82]

Wallace *et al.* applied resonance Raman spectroscopy to the investigation of the reaction of the tetrathiafulvalene-TCNQ electrode in an aqueous solution, and found that the oxidative decomposition of the electrode resulted in the formation of neutral TCNQ.[83] The reduction and oxidation of methylviologen (MV) was investigated on silver and *p*-type β-ZnP$_2$ electrodes with resonance Raman measurements.[84] The formation of dissolved MV$^{\cdot+}$ radical ion was observed during the reduction reaction of MV^{2+} at the Ag electrodes. The adsorbed radical ion was also detected during the reaction when a roughened electrode was used in dilute solutions. The oxidized colorless component MV^{2+} also could be detected on the Ag electrode roughened electrochemically, because of the surface enhanced Raman effect. On the β-ZnP$_2$ electrodes, MV$^{\cdot+}$ ions were formed by photoelectrochemical reduction of MV^{2+} under the incidence of the laser beam for the Raman excitation. The reduction was monitored as a function of applied voltage using the resonance Raman measurement.

Akins *et al.* observed resonance and FT-Raman spectra of 1,1′-diethyl-

4,4'-cyanine iodide adsorbed on a smooth silver electrode using visible and near-infrared excitations, respectively.[85] The obtained spectra imply that the cyanine dye is present on the electrode mainly in the form of polycrystallite and aggregate molecules; whereas the dye is adsorbed prevalently as polycrystallite at high pH, in the low-pH system the dye molecules aggregate in its protonated form. The FT-Raman observation showed that the intensity of Raman bands increased in the low-pH system, and the intensity increase was attributed to "aggregation enhancement," which was associated with an increase in polar-izabilty as a result of the aggregate formation.[86]

The observation of resonance Raman spectra of colored species is thus helpful for the investigation of the behavior of the species on electrodes. However, colored species absorb both the exciting laser beam and the Raman scattered light, resulting in a decrease in the observed intensity of the Raman bands. Consequently, electrolytic cells, in which the thickness of the solution is less than 1 mm, are generally used. Furthermore, the continuous irradiation of the colored species with an intense laser beam often results in the decomposition of the sample molecules. When a colored species is formed from a colorless one in an electrochemical reaction, however, these drawbacks can be removed by the use of an oscillating electrode potential (the cyclic double potential step) or potential modulation Raman spectroscopy described below, and the colored species can be irradiated without problem by a laser beam whose wavelength lies within the absorption band of the sample species.[81,87]

In the potential oscillation measurement, the electrode potential is modulated with a square wave AC signal. Let us suppose that a soluble colored species is produced from a colorless species in the solution in an oxidation reaction on the electrode and that the reaction is electrochemically reversible. When the oxidation potential is applied to the electrode, the colored species is produced at the electrode and begins to diffuse into the bulk solution. However, the potential is shifted after a very short time interval to the cathodic value so that the colored species is reduced to the uncolored one. The colored species therefore, is virtually confined in a thin layer on the electrode. As a result, the absorption of the exciting beam and the Raman scattered light is minimized. The decomposition of the colored species by the laser irradiation is also minimized because it is converted into the colorless species after a short time interval. In addition, since the obtained Raman spectrum provides information about species freshly formed at the electrode, the spectrum of a metastable reaction intermediate may be obtained in the potential oscillation measurement.

Figure 10 shows the spectrum of compounds formed in an oxidation reaction on a Pt electrode in an acidic solution of 3,3'-dimethoxybenzidine.[87] The asterisks in the figure designate Raman bands attributable to the fully

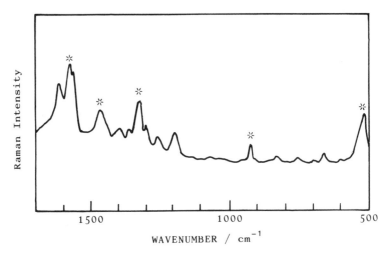

Figure 10. Resonance Raman spectrum of species formed in electro-oxidation of 3,3'-dimeth-oxybenzidine measured with the potential oscillation method. Electrode potential was oscillating at a frequency 10 Hz between 0.45 and 0.55 V (versus SCE). Concentration of the reactant, 1×10^{-3} M; wavelength of the exciting laser light, 514.5 nm (adapted from Ref. 87).

oxidized monomer of quinone diimine structure. These bands decrease rapidly in intensity with time after the reaction, because the concentration of the monomer decreases rapidly by a reaction called a reproportionation.[88]

When the colored species formed in an electrochemical reaction is not soluble and is adsorbed or grows in a thin film on the electrode, the potential modulation method[89,90] is applicable to the *in situ* observation of Raman spectra of the colored species. In the potential modulation method, the oscillation of potential is combined with the synchronous single photon counting technique.[91] In the potential modulation Raman measurement, the difference in the spectra of species at two preset potentials is directly recorded, as is the case in electrochemical modulation infrared spectroscopy. Raman bands of species immune to the potential change, such as the supporting electrolyte and the solvent molecules, the tail of the Rayleigh band, and the fluorescence from the immune species, are eliminated from the obtained spectrum. When a colored species is produced from a colorless species in the electrochemical reaction, the resonance Raman spectrum of the colored species is obtained with the potential modulation Raman measurement, in contrast to the electrochemical modulation IR measurement, because the signals from the colorless species are negligible in intensity. This is also the case when the reaction product and the reactant molecule are differently colored, provided the wavelength of the incident laser light is chosen to excite only the resonance scattering of the former. On the other hand, if both the species present on the electrode at two distinct potentials yield the SERS, the net difference of the two SER spectra is measured with the modulation method.

Figure 11b shows a resonance Raman spectrum of mono-cation radical of n-heptylviologen ($HV^{\cdot+}$) deposited on a smooth silver electrode in the reduction reaction of colorless heptylviologen dication (HV^{2+}).[92] Signals from the solvent (H_2O), the supporting electrolyte ($H_2PO_4^-$), and the Rayleigh band tail are eliminated from the modulated spectrum. This modulation technique can be applied to the *in situ* observation of electrochemical reactions on semiconductor electrodes as well. The resonance Raman spectrum of an $HV^{\cdot+}$ thin film deposited on an In_2O_3 electrode was obtained using the modulation method.

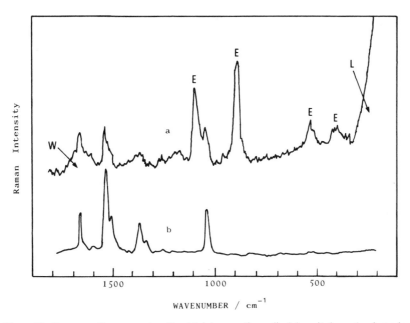

WAVENUMBER / cm^{-1}

Figure 11. Resonance Raman spectra of heptylviologen cation radical deposited on a Ag electrode. Raman bands of the solvent and $H_2PO_4^-$ ion and the tail of the Rayleigh band are designated with W, E and L, respectively. Electrode potential was oscillating at a frequency of 10 Hz between −0.3 and −0.7 V (versus SCE). (a) Potential oscillation measurement; (b) potential modulation measurement. Wavelength of exciting light, 514.5 nm; concentration of HV^{2+}, 1×10^{-3} M (adapted from Ref. 94).

The Raman band at 1510 cm^{-1}, which was assignable to the dimer of HV$^{·+}$ and appeared in the spectra of the films on Ag and Pt electrodes, was absent in the spectrum of the film on the In$_2$O$_3$ electrode.[93]

Figure 12 shows a potential modulation Raman spectrum of pyridine adsorbed on a roughened silver electrode.[89] This spectrum shows features on each side of the base line, because the difference of two SER spectra has been recorded. The change in wavelength of the exciting laser beam from 514.5 nm to 607.4 nm resulted in a remarkable decrease in intensity of the Raman bands, showing the enfeeblement of the enhancement, but the bands recovered their intensity when the electrode potential was modulated in the slightly more anodic range of −0.5 ± 0.1 V (versus SCE). This suggests that the enhancement may arise from a resonance mechanism. The potential modulation Raman measurement therefore, may be used for the investigation of the mechanism of the "chemical" enhancement in SERS.

Figure 13 is a potential modulation SER spectrum of water adsorbed on a roughened silver electrode.[94] Fleischmann *et al.*[95] obtained a similar spectrum using a thin layer cell and a highly concentrated halide solution. The potential modulation method does not require the use of the thin layer cell. The

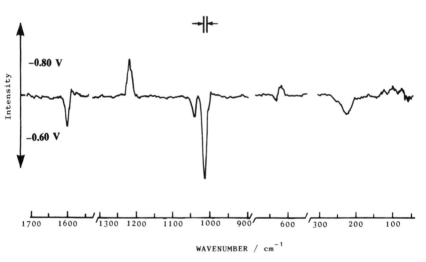

Figure 12. Potential modulation Raman spectrum of pyridine adsorbed on a roughened silver electrode. The electrode potential was modulated between −0.6 and −0.8 V (versus SCE) at a frequency of 20 Hz. Concentration of pyridine, 0.05 *M* [Reprinted from W. Suëtaka and M. Ohsawa, *Appl. Surf. Sci.* **3**, 188 (1979)].

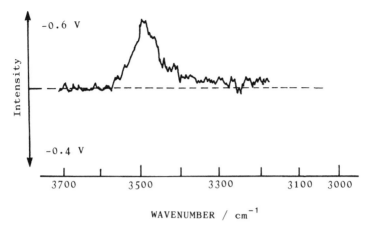

Figure 13. Potential modulation Raman spectrum of water adsorbed on a roughened silver electrode in an aqueous solution containing 0.1 M KCl. The electrode potential was modulated at a frequency of 10 Hz between −0.4 and −0.6 V (versus SCE). Wavelength of the exciting laser light, 514.5 nm (adapted from Ref. 94).

Raman band is narrow in width and shifted to a higher wavenumber in comparison to that of bulk water, showing the formation of weak intermolecular hydrogen bonds.

Several techniques are used for the time-resolved observation of Raman spectra of species on electrode surfaces. Clarke *et al.* observed the intensity change at a fixed wavenumber after a stepwise change in the electrode potential.[96] Jeanmaire *et al.* investigated diffusion controlled electrode reactions in observing at a fixed wavenumber the Raman intensity versus time transients following the single-shot double-potential step.[81,97] Both the potential oscillation method combined with the photon counting technique and the potential modulation method can be used to follow the rapid change in the spectrum of surface species with time. In the time-resolved observation, a short sampling time ranging from 1 to 20 ms is used, and a series of spectra is recorded by changing the delay time, which is the time lag between the potential step and the opening of the electronic gate of the photon counter. Figure 14 is an example of the time-resolved Raman spectra of species on an electrode surface measured with the potential oscillation method.[98] This figure shows the growth of an $HV^{\cdot+}$ film on a Pt electrode in an aqueous solution containing HV^{2+} (1×10^{-3} M) and KBr (0.3 M). The Raman bands at 1510 and 1530 cm^{-1} are assignable to the dimer of the cation radical, $(HV^{\cdot+})_2$, and the bromide of the cation, respectively. Whereas the band of the dimer remains constant in in-

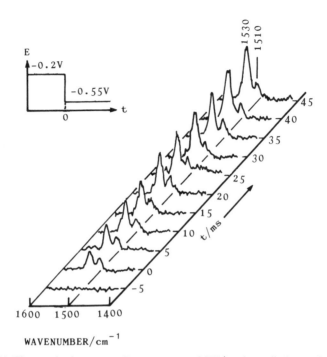

WAVENUMBER/cm^{-1}

Figure 14. Time-resolved resonance Raman spectra of HV$^{\cdot+}$ cation radical on a Pt electrode measured by changing the delay time after the potential step from −0.2 to −0.55 V (versus Ag/AgCl). Time (t) is referenced to the potential change. Sampling time, 5 ms; excitation wavelength, 514.5 nm [Reprinted from M. Osawa and W. Suëtaka, *J. Electroanal. Chem.* **270**, 261 (1989)].

tensity independent of the reduction duration, the intensity of the bromide band continues to increase with time, corresponding to the diffusion-controlled growth of the surface film.

Pulsed laser irradiation is used for the time-resolved Raman observation of charge-transfer reactions on semiconductor particles. In this observation, e^- and h^+ are created by a single laser pulse above the semiconductor band gap. After a preset time lag, a resonance Raman spectrum is generated by a second pulse at a frequency below the band gap. This method was applied to investigate the hole transfer to SCN$^-$ on TiO$_2$ and the electron transfer to MV^{2+} on TiO$_2$ and CdS.[99]

The vibrational spectra of corrosion products on metal surfaces formed in dry and wet corrosion are generally measured with infrared spectroscopy, because the sensitivity of IR spectroscopy is higher than that of unenhanced Raman scattering and the corrosion products are often colored or fluorescent. Unenhanced Raman spectroscopy therefore, is principally used for the *in situ* observation of the corrosion of metal in aqueous media.

Passive films grown on iron electrodes roughened with oxidation-reduction cycles was observed *in situ* with Raman spectroscopy. The obtained spectra were analyzed in comparison with those of thermally grown oxides. Raman bands of magnetite were located for the film on iron anodically oxidized in a KOH solution. The *in situ* observation of passive films on iron formed in sulfuric acid solution, however, was strongly hampered by the signals from the acid. *Ex situ* observation of the passive films was then performed and the spectra obtained showed the presence of Fe_2O_3 and Fe_3O_4.[100] The anodic corrosion films of Pb, Ag, Cu, Ni, and Co were also observed *in situ* with Raman spectroscopy.[101–103] Faultless passive films were formed on iron electrodes in basic phosphate solutions. The Raman measurement of the passive films showed the presence of Fe_3O_4 and phosphate ion incorporated into the film. On the other hand, dissolution of the electrode occurred in acidic phosphate solutions and vivianite, $Fe_3 (PO_4)_2 \cdot 8H_2O$, was found as the main component of the film formed on iron.[104]

Halide ions are well known to cause the breakdown of the passivity of metals and alloys in aqueous media. Since thiocyanate ion is a pseudohalide, it may cause the breakdown of the passivity, and it has a stretching Raman band that changes significantly in frequency depending upon the bonding type.[105] Its behavior has been studied with the Raman measurement to better understand the breakdown phenomena of passivity. Thiocyanate ion brought about the breakdown of the passivity on iron, but in contrast, copper maintained its passivity in the presence of the ion. The Raman measurement of iron and copper surfaces in a borate buffer solution containing thiocyanate ion has shown that a CuSCN film is formed on Cu and that Fe(II) and Fe(III) thiocyanate, probably incorporated in oxides or oxyhydroxides, is present on iron.[106]

In situ observation of Raman spectra of metal surfaces in high temperature/high pressure aqueous media should provide information valuable for the fundamental understanding of the mechanism of stress corrosion cracking, which is a serious problem in the cooling systems of power plants and nuclear reactors. Melendres *et al.*[77,107] have studied the corrosion and passivation of lead and nickel electrodes in a dilute Na_2SO_4 solution at high temperature by means of Raman spectroscopy and cyclic voltammetry. Nickel does not maintain its passivity and dissolves in solution above 373 K. The Raman observation in the secondary passivation region (about 0.55 V versus Ag/Ag_2SO_4) showed the presence of "hydrous NiO_2," while no Raman band was found in the primary passivation region because of the insufficient scattering volume. Passive films were formed on lead and their composition was deduced from the analysis of the obtained Raman spectra. At high temperatures, $PbOPbSO_4$ and $PbSO_4$ were found in the surface film at open circuit and the latter took a more effective part as passivator than the former.

Stainless steel type 304 is used widely in the piping of nuclear reactors, but the sensitized steel is susceptible to intergranular stress corrosion cracking (IGSCC) because of the formation of a chromium-depleted layer. Recently published potential-pH (Pourbaix) diagrams show that $FeCr_2O_4$ is the stable oxide phase in the conditions where IGSCC does not occur, while Fe_2O_3 is the stable oxide when IGSCC takes place.[108] Employing Raman and x ray photoelectron spectroscopies, Maroni *et al.*[109] have investigated the surface film formed on sensitized type 304 SS steel exposed to water at 562 K under various pHs and applied potentials. The results indicate that the above-mentioned Pourbaix diagrams[108] give an accurate description of the phase stability domains in the pH range from 3.0 through 7.0.

Since the electrochemical double layer on the electrodes can be emersed intact from solution,[110] Pemberton and Sobocinski[111] measured the Raman spectra from Ag electrodes emersed from alcohol electrolyte solutions for the investigation of the behavior of solvent molecules on the electrodes. The spectra obtained from the electrodes emersed at various electrode potentials were compared with SER spectra of the electrochemically roughened Ag electrodes held at the same potentials. A thin layer cell was used in the SERS measurement for minimizing the Raman signals from the bulk solution. The spectra of solvent molecules on the emersed electrodes were in agreement with the SER spectra observed *in situ* in alcohol electrolyte solutions, showing that the orientation of the solvent molecules changed depending upon the electrode potential. The results obtained suggest that the orientation of solvent molecules on the electrodes, as well as their bonding to the electrode surface, remain unperturbed by the emersion process.

References

1. T. R. Gilson and P. J. Hendra, *Laser Raman Spectroscopy*, Wiley, New York (1970).
2. J. A. Koningstein, *Introduction to the Theory of the Raman Effect*, Reidel, Dordrecht, Holland (1972).
3. D. A. Long, *Raman Spectroscopy*, McGraw-Hill, New York (1977).

4. J. G. Grasselli, M. K. Snavely, and B. J. Bulkin, *Chemical Applications of Raman Spectroscopy,* Wiley, New York (1981).
5. D. P. Strommen and K. Nakamoto, *Laboratory Raman Spectroscopy,* Wiley, New York (1985).
6. W. Krasser and A. J. Renouprez, *J. Raman Spectrosc.* **8,** 92 (1979).
7. P. P. Yaney and R. J. Becker, *Appl. Surf. Sci.* **4,** 356 (1980).
8. J. Tang and A. C. Albrecht, in: *Raman Spectroscopy: Theory and Practice,* Vol. 2 (H. A. Szymanski, ed.), Plenum, New York (1970), p. 33.
9. R. J. H. Clark, in: *Advances in Infrared and Raman Spectroscopy,* Vol. 1 (R. J. H. Clark and R. E. Hester, eds.), Heyden, London (1976).
10. H. Yamada, *Appl. Spectrosc. Rev.* **17,** 227 (1981).
11. A. Campion, in: *Vibrational Spectroscopy of Molecules on Surfaces* (J. T. Yates, Jr. and T. E. Madey, eds.), Plenum, New York (1987), p. 345.
12. A. Campion, *J. Electron Spectrosc. Relat. Phenom.* **54/55,** 877 (1990).
13. J. A. Koningstein and B. F. Gächter, *J. Opt. Soc. Am.* **63,** 892 (1973).
14. H. Yamada and Y. Yamamoto, *J. Raman Spectrosc.* **9,** 401 (1980).
15. P. J. Hendra, ed., *Spectrochim. Acta* **46A** (1990).
16. R. G. Greenler and T. L. Slager, *Spectrochim. Acta* **29A,** 193 (1973).
17. D. R. Mullins and A. Campion, *J. Phys. Chem.* **88,** 8 (1984).
18. M. Ohsawa, W. Kusakari, and W. Suëtaka, *Spectrochim. Acta* **36A,** 389 (1980).
19. Z. Lenac and M. S. Tomaš, *J. Raman Spectrosc.* **22,** 831 (1991).
20. M. Moskovits, *J. Chem. Phys.* **77,** 4408 (1982).
21. V. M. Hallmark and A. Campion, *J. Chem. Phys.* **84,** 2933, 2942 (1986).
22. T. Ikeshoji, Y. Ono, and T. Mizuno, *Appl Opt.* **12,** 2236 (1973).
23. K. H. Redel, D. Spenkuch, and G. R. Wessler, *Opt. Spectrosc.* **38,** 649 (1975).
24. M. Fujihara and T. Osa, *J. Am. Chem. Soc.* **98,** 7850 (1976).
25. W. Carious and O. Schrötter, *Z. Phys. Chem.* (Leipzig) **262,** 711 (1981).
26. J. Cipriani, S. Racine, R. Dupeyrat, H. Hasmonay, M. Dupeyrat, Y. Levy, and C. Imbert, *Opt. Comm* **11,** 70 (1974).
27. J. F. Rabolt, R. Santo, and J. D. Swalen, *Appl. Spectrosc.* **33,** 549 (1979).
28. R. Iwamoto, M. Miya, K. Ohta, and S. Mima, *J. Am. Chem. Soc.* **102,** 1212 (1980); *J. Chem. Phys.* **74,** 4780 (1981).
29. T. Takenaka and T. Nakanaga, *J. Phys. Chem.* **80,** 475 (1976).
30. G. S. Agarwal, *Phys. Rev.* **A12,** 1475 (1975).
31. W. Lukosz and R. E. Kunz, *J. Opt. Soc. Am.* **67,** 1607, 1615 (1977).
32. W. Lukosz, *J. Opt. Soc. Am.* **69,** 1495 (1979); **71,** 744 (1981).
33. W. Lukosz and R. E. Kunz, *Opt. Commn.* **31,** 251 (1979).
34. G. Mattei, B. Fornari, and M. Pagannone, *Solid State Comm.* **36,** 309 (1980).
35. M. Ohsawa, K. Hashima, and W. Suëtaka, *Appl. Surf. Sci.* **20,** 109 (1984).
36. J. F. Rabolt and J. D. Swalen, in: *Spectroscopy of Surfaces* (R. J. H. Clarke and R. E. Hester, eds.), Wiley, New York (1988), p. 1.
37. Y. Levy, C. Imbert, J. Cipriani, S. Racine, and R. Dupeyrat, *Opt. Comm.* **11,** 66 (1974).
38. J. F. Rabolt, R. Santo, and J. D. Swalen, *Appl. Spectrosc.* **34,** 517 (1980).
39. J. F. Rabolt, N. E. Schlotter, and J. D. Swalen, *J. Phys. Chem.* **85,** 4141 (1981).
40. C. G. Zimba, V. M. Hallmark, S. Turrell, J. D. Swalen, and J. F. Rabolt, *J. Phys. Chem.* **94,** 939 (1990).
41. C. G. Zimba and J. F. Rabolt, *Thin Solid Films* **206,** 388 (1991).

42. G. I. Stegeman, R. Fortenberry, R. Moshrefzadeh, W. H. Hetherington III, N. E. Van Wyck, and J. E. Sipe, *Opt. Lett.* **8**, 295 (1983).
43. W. H. Hetherington III, Z. Z. Ho, E. W. Koenig, G. I. Stegeman, and R. Fortenberry, *Chem. Phys. Lett.* **128**, 150 (1986).
44. G. Karagounis and R. Issa, *Z. Elektrochem.* **66**, 874 (1962).
45. R. O. Kagel, *J. Phys. Chem.* **74**, 4518 (1970).
46. T. A. Egerton, A. H. Hardin, Y. Kozirovsky, and N. Sheppard, *J. Catal.* **32**, 343 (1974).
47. P. J. Hendra, J. R. Horder, and E. J. Loader, *J. Chem. Soc.* **1766** (1971); P. J. Hendra, I. D. M. Turner, E. J. Loader, and M. Stacey, *J. Phys. Chem.* **78**, 300 (1974).
48. M. L. Howe, K. L. Watters, and R. G. Greenler, *J. Phys. Chem.* **80**, 382 (1976).
49. J. M. Stencel and E. B. Bradley, *Spectrosc. Lett.* **11**, 563 (1978).
50. J. M. Stencel, D. M. Noland, E. B. Bradley, and C. A. Frenzel, *Rev. Sci. Instrum.* **49**, 1163 (1978).
51. J. M. Stencel and E. B. Bradley, *J. Raman Spectrosc.* **8**, 203 (1979).
52. H. A. Marzouk, K. A. Arunkumar, and E. B. Bradley, *Surf. Sci.* **147**, 477 (1984).
53. A. Campion, J. K. Brown, and V. M. Grizzle, *Surf. Sci.* **115**, L153 (1982).
54. A. Campion, and D. R. Mullins, *Surf. Sci.* **158**, 263 (1985).
55. D. Harradine and A. Campion, *Chem. Phys. Lett.* **135**, 501 (1987).
56. C. Shannon and A. Campion, *Surf. Sci.* **227**, 219 (1990).
57. R. N. Zitter, *Surf. Sci.* **28**, 335 (1971).
58. J. A. Cape, L. G. Hale, and W. E. Tennant, *Surf. Sci.* **62**, 639 (1977).
59. R. Tsu and S. S. Jha, *J. Phys.* (Paris) **41**, C4-25 (1980).
60. H. J. Stolz and A. Abstreiter, *Solid State Comm.* **36**, 857 (1980).
61. G. P. Schwartz, B. Schwartz, D. DiStefano, G. J. Gualtieri, and J. E. Griffiths, *Appl. Phys. Lett.* **34**, 205 (1979).
62. Y. Dai, J. S. Swinnea, H. Steinfink, J. B. Goodenough, and A. Campion, *J. Am. Chem. Soc.* **109**, 5291 (1987).
63. G. Mattei, B. Fornari, M. Pagannone, and L. Mattioli, *J. Phys.* (Paris) **45**, C5-249 (1984); *Solid State Comm.* **44**, 1495 (1982).
64. Y. Tian, J. Umemura, T. Takenaka, and T. Kunitake, *Langmuir* **4**, 1064 (1988).
65. D. R. Miller, O. H. Hahn, and P. W. Bohn, *Appl. Spectrosc.* **41**, 245 (1987).
66. J. P. Rabe, J. D. Swalen, and J. F. Rabolt, *J. Chem. Phys.* **86**, 1601 (1987).
67. S. S. Perry and A. Campion, *Surf. Sci.* **234**, L275 (1990).
68. S. S. Perry and A. Campion, *J. Electron Spectrosc. Relat. Phenom.* **54/55**, 933 (1990).
69. S. S. Perry and A. Campion, *Surf. Sci.* **259**, 207 (1991).
70. M. Fleischmann, P. J. Hendra, and A. J. McQuillan, *J. Chem. Soc. Chem. Comm.* **80** (1973).
71. M. Fleischmann, P. J. Hendra, and A. J. McQuillan, *Chem. Phys. Lett.* **26**, 163 (1974).
72. E. S. Reid, R. P. Cooney, P. J. Hendra, and M. Fleischmann, *J. Electroanal. Chem.* **80**, 405 (1977).
73. M. Pourbaix, *Atlas of Electrochemical Equilibria in Aqueous Solutions,* Pergamon, Oxford (1966).
74. B. H. Loo, *J. Phys. Chem.* **86**, 433 (1982).
75. B. H. Loo and Y. G. Lee, *Appl. Surf. Sci.* **18**, 345 (1984).
76. C. Shannon and A. Campion, *J. Phys. Chem.* **92**, 1385 (1988).
77. C. A. Melendres, J. J. McMahon, and W. Ruther, *J. Electroanal. Chem.* **208**, 175 (1986); *Proceedings of the Tenth International Congress on Metallic Corrosion* **4**, 3481 (1987).
78. C. A. Melendres, *J. Electroanal. Chem.* **286**, 273 (1990).

79. W. P. Acker, B. Tip, D. H. Leach, and R. K. Chang, *J. Appl. Phys.* **64**, 2263 (1988).
80. D. K. Veirs, V. K. F. Chia, and G. M. Rosenblatt, *Appl. Opt.* **26**, 3530 (1987).
81. D. L. Jeanmaire, M. R. Suchanski, and R. P. Van Duyne, *J. Am. Chem. Soc.* **97**, 1699 (1975).
82. M. R. Suchanski and R. P. Van Duyne, *J. Am. Chem. Soc.* **98**, 250 (1976).
83. W. L. Wallace, C. D. Jaeger, and A. J. Bard, *J. Am. Chem Soc.* **101**, 4840 (1979).
84. I. Blatter-Mörke, H. von Känel, and P. Wachter, *J. Phys. Chem.* **91**, 663 (1987).
85. D. L. Akins, J. W. Macklin, and H-R. Zhu, *J. Phys. Chem.* **96** 4515 (1992).
86. D. L. Akins, J. W. Macklin, and H-R. Zhu, *J. Phys. Chem.* **95**, 793 (1991).
87. M. Ohsawa and W. Suëtaka, *Ext. Abst. 26th Intern. Cong. Pure Appl. Chem.* **1**, 152 (1977).
88. N. Winograd and T. Kuwana, *J. Am. Chem. Soc.* **93**, 4343 (1971).
89. W. Suëtaka and M. Ohsawa, *Appl. Surf. Sci.* **3**, 118 (1979).
90. R. P. Van Duyne, in: *Chemical and Biochemical Applications of Lasers,* Vol. 4 (C. B. Moore, ed.), Academic, New York (1979), p. 101.
91. F. T. Arecchi, E. Gatti, and A. Sona, *Rev. Sci. Instrum.* **37**, 942 (1966).
92. M. Ohsawa, K. Nishijima, and W. Suëtaka, *Surf. Sci.* **104**, 270 (1980).
93. N. Yagi and W. Suëtaka, unpublished data.
94. M. Ohsawa, D. E. Thesis, Tohoku University (Sendai) 1984.
95. M. Fleischmann, P. J. Hendra, I. R. Hill, and M. E. Pemble, *J. Electroanal. Chem.* **117**, 243 (1981).
96. J. S. Clarke, A. T. Kuhn, and W. J. Orville-Thomas, *Electroanal. Chem. Interfacial Electrochem.* **54**, 253 (1974).
97. D. L. Jeanmaire and R. P. Van Duyne, *J. Electroanal. Chem.* **60**, 235 (1974).
98. M. Osawa and W. Suëtaka, *J. Electroanal. Chem.* **270**, 261 (1989).
99. R. Rosetti, S. M. Beck, and L. E. Brus, *J. Am. Chem. Soc.* **104**, 7322 (1982); *J. Am. Chem. Soc.* **106**, 980 (1984).
100. M. Froelicher, A. Hugot-Le Goff, C. Pallotta, R. Dupeyrat, and M. Masson, in: *Passivity of Metals and Semiconductors* (M. Froment, ed.), Elsevier, Amsterdam (1983), p. 101.
101. R. J. Thibeau, C. W. Brown, A. Z. Goldfarb, and R. H. Heidensbach, *J. Electrochem. Soc.* **127**, 37 (1980).
102. R. Kötz and E. Yeager, *J. Electroanal. Chem.* **111**, 105 (1980).
103. C. A. Melendres and S. Xu, *J. Electroanal. Chem.* **162**, 343 (1984); *J. Electrochem. Soc.* **131**, 2239 (1984).
104. C. A. Melendres, N. Camillone, and T. Tipton, *Electrochim. Acta* **34**, 281 (1989).
105. K. Nakamoto, *Infrared and Raman Spectra of Inorganic and Coordination Compounds,* 4th Ed., Wiley-Interscience, New York (1986).
106. C. A. Melendres, T. J. O'Leary and J. Solis, *Electrochim. Acta* **36**, 505 (1991).
107. J. J. McMahon, W. Ruther, and C. A. Melendres, *J. Electrochem. Soc.* **135**, 557 (1988).
108. D. Cubicciotti, *J. Nucl. Mater.* **167**, 241 (1989).
109. V. A. Maroni, C. A. Melendres, T. F. Kassner, and S. Siegel, *J. Nucl. Mater.* **172**, 13 (1990).
110. G. J. Hansen and W. N. Hansen, *J. Electroanal. Chem.* **150**, 193 (1983).
111. J. E. Pemberton and R. L. Sobocinski, *J. Electroanal. Chem.* **318**, 157 (1991).

6

Surface Enhanced Raman Scattering

1. Introduction

Surface enhanced Raman scattering (SERS)[1-9] was one of the most exciting subjects in surface science for about 10 years after its discovery. It is now widely used as an analytical tool in applied science because of its high sensitivity, although the full consensus on this enhancement mechanism has not yet been reached.

The discovery and subsequent ground-breaking work of SERS were done by spectroelectrochemists, because Raman spectroscopy was believed to be a potential tool for the investigation of adsorbed species and their behavior on electrodes. The sensitivity of normal Raman scattering, however, is not sufficient for easy observation of surface species, as discussed in Chapter 5. The intensity of Raman scattered light increases with the frequency of the scattered light, the strength of the exciting electric field, and the components of the derived polarizability tensor of the vibrational modes as well as the scattering volume, as described in the preceding chapter. It was quite natural, therefore, that Fleischmann *et al.* used electrochemically roughened Ag electrodes in their Raman observation of pyridine adsorbed on the electrodes on purpose to increase the Raman scattering volume.[10] The surprisingly strong Raman signals of pyridine they observed caught the attention of Van Duyne, Creighton, and their coworkers. They repeated the Raman measurement using the same system and found that the Raman scattering cross section of adsorbed pyridine was enormously increased.[11,12] In the measurement using a solution of 10^{-2} M pyridine, for example, only faint signals from bulk solution were obtained prior to the electrochemical roughening of the Ag electrode, but anomously intense

Raman signals of adsorbed pyridine were observed after a single oxidation-reduction cycle.[12] Based on this measurement, Albrecht and Creighton estimated an enhancement of the Raman scattering cross section of about 10^5 times for pyridine adsorbed on the roughened Ag electrode.

This enormous 10^5–10^6 enhancement of Raman scattering intensity strongly stimulated theoretical work on the mechanism of the enhancement. Albrecht and Creighton[12] discussed the possibility of resonant Raman scattering through interaction with the surface plasmon, the mechanism proposed by Philpott.[13] On the other hand, Van Duyne suggested the possibility of the enhancement arising from the large electric field present in the outer Helmholz layer.[14] Moskovits proposed the resonant excitation of the transverse collective electron resonance in metal bumps on the surface of electrochemically roughened Ag electrodes for the origin of the enhanced Raman scattering.[15] After these early works, numerous interesting mechanisms were proposed to explain the enhanced scattering.

Presently two mechanisms, electromagnetic and "chemical" enhancement, are generally accepted as the principal causes of the enhancement. There is no doubt about the existence of the electromagnetic (EM) enhancement associated with surface plasmon polariton (SPP), the plasma resonance by metal particles and the collective electron resonance, as discussed in Chapter 3. The strong EM field associated with the SPP and the collective electron resonance gives rise to the enhancement of absorption, fluorescence, and photochemical processes, in addition to Raman scattering, resulting in the birth of new techniques included under the general name of "surface enhanced spectroscopy."[4,16] Furthermore, the strong EM field is utilized in the surface plasmon sensor for measuring the refractive index and the absorptions of thin layers on solid surfaces.[17,18] The surface plasmon microscope also takes advantage of the strong surface EM field for the measurement of the two-dimensional distribution of the thickness[19,20] or the refractive index[21] of thin films formed on metal substrates.

A kind of resonance Raman scattering, in which a charge transfer between adsorbates and metal surfaces is involved, is proposed for the so-called "chemical" enhancement.[1,9] There still remains, however, controversy about the detailed mechanism of adsorption-induced resonance Raman scattering.

Since the "chemical" enhancement is specific to chemical species capable of forming adequate adsorption-induced states, the surface of interest cannot be surveyed with SERS when only the "chemical" mechanism is functioning on the surface. Furthermore, the extent of the contribution to the Raman signals from each enhancement mechanism varies depending upon the system, and no internal standard can be used for the estimation of the enhancement factor due to the "chemical" mechanism. The measurement of SERS is not appropriate, therefore, for the quantitative analysis of surface species whenever the "chemical" mechanism is involved in the enhancement. In spite of these difficulties, the highly sensitive SERS measurement is widely applied to the investigation of a variety of surface phenomena, such as adsorption and reaction on electrode surfaces, metallic corrosion and its inhibition, and the structure and molecular orientation of thin polymer and Langmuir–Blodgett films.

Fourier transform-Raman spectroscopy is applied to practical problems as a fluorescence-free technique. However, the sensitivity of this technique is insufficient for surface studies, because a near-infrared laser is used as the exciting source. Resonance Raman enhancement can be used to improve the low sensitivity, but most materials have no absorption in the near-infrared region. Surface enhanced Raman scattering is therefore utilized for the enhancement of the signal-to-noise ratio of FT-Raman spectroscopy.[22–24]

2. Mechanism of Enhancement

2.1. Electromagnetic (EM) Mechanisms

Surface enhanced Raman spectra were first obtained using electrochemically roughened Ag electrodes. The rough surfaces of gold and copper also give rise to SER spectra when incident light of a wavelength outside the electronic interband transitions ($\lambda > 570$ nm) is used. The rough surface may be considered as a combination of various gratings. Consequently, when a beam of p-polarized light is incident at an appropriate angle on a grating of metals such as Ag, Au, or Cu in the wavelength region where they show free electron metal-like behavior, SPP can be generated on the metal surface, as described in Section 2.2 of Chapter 3. The EM field of SPP is stronger in intensity than the incident light, and therefore, the Raman-shifted light is increased in intensity. In addition, the Raman-shifted light can excite SPP, though to a lesser extent, thus giving rise to further enhancement of the Raman signals.[4,25]

The enhancement of Raman signals through the discrete Fourier component of the grating was first observed by Tsang et al.[26] Figure 1 shows a typical result of the grating-assisted enhancement obtained by Girlando et al. using holographic silver gratings coated with a thin polystyrene film.[27] The reflectivity of p-polarized light shows a remarkable dip near an angle of incidence of 4°. The dip indicates the grating-assisted absorption of light by the surface plasmon. The Raman spectrum of the polystyrene film (~10 nm thick) deposited on the grating shows the intensity maximum at the same angle of incidence as can be seen in Fig. 1b. On the other hand, the incidence of s-polarized light gives rise to neither the reflectivity dip nor the intensity maximum of the Raman scattered light. This result clearly indicates the intensity increase of the Raman signal by the grating-assisted SPP excitation. The enhancement of Raman signals through the surface roughness-assisted SPP excitation was also demonstrated in reverse-ATR measurements made by Hayashi.[28] The enhancement factor by the surface roughness-assisted SPP excitation changes depending upon the depth of the trough of the rough sur-

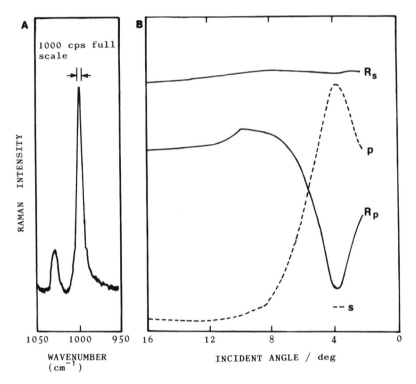

Figure 1. (a) Raman spectrum of a 10 nm polystyrene film excited with p-polarized laser light incident at 4° on the grating; (b) reflectivity of the silver grating for p- and s-polarized light (solid line) and the intensity of the Raman line at 1000 cm⁻¹ (assignable to the totally symmetric vibration of benzene ring) as a function of the incident angle (dashed line). Raman scattering was excited by p- and s-polarized laser light at 530.9 nm, grating: Λ = 450.7 nm, H = 16.3 nm [Reprinted from A. Girlando, M. R. Philpott, D. Heitmann, J. D. Swalen, and R. Santo, *J. Chem. Phys.* **72**, 5187 (1980)].

faces, and a maximum enhancement of 1.5×10^4 was obtained at a depth of ~10nm for a moderately roughened surface.[29]

The surface plasmon polariton can also be excited in the Otto- and Kretschmann-ATR configurations, as described in Section 2.2 of Chapter 3. Since, in the visible region, coinage metals are not so strongly absorbing as in the infrared, the Kretschmann configuration is generally used for the excitation of SPP. Prior to the discovery of the enhancement of the Raman cross section of pyridine molecules adsorbed on Ag electrodes, Chen et al.[30] predicted theoretically the enhancement of Raman signals in the ATR configuration in the presence of a thin silver film, and they observed experimentally the enhancement of the Raman signals from a thin organic film coating over the metal layer. The measurement of enhanced Raman signals by the ATR-assisted SPP excitation were also performed by Pettinger et al.[31] and Dornhaus et al.[32]

Figures 2 and 3 show the results obtained by Dornhaus et al.[32] The bottom spectrum in Fig. 2 was obtained from isonicotinic acid molecules adsorbed on a 57 nm Ag film on an ATR prism with the incidence of p-polarized light at the resonance angle (θ_p) for the SPP excitation. Figure 3 shows the decrease in intensity of the Raman band at 1600 cm^{-1} of p-nitrobenzoic acid adsorbed on a 57 nm Ag film as the angle of incidence of the exciting radiation is increased. The intensity of the band decreases steeply when the incident angle increases from the resonance angle. These results indicate that the Raman signals are enhanced by the strong EM field of SPP. Ushioda and Sasaki[33] measured the intensity of Raman bands of methanol molecules on an evaporated 40 nm Ag film. The exciting laser beam was incident through the prism and the Raman scattered light also was collected through the prism. The observed intensity changed sharply with the incident and collected angles, showing SPP was excited by the incident laser beam as well as by the Raman scattered light. An enhancement factor of about 4×10^4 was estimated in this work. The facts hitherto mentioned demonstrate that SPP plays an important role in SERS. Because the field of SPP decays exponentially into the media, the enhancement by SPP is long-range.

The top spectrum in Fig. 2 was obtained from isonicotinic acid adsorbed on a 5 nm Ag island film with the light incident at the critical angle of total internal reflection (θ_c).[32] The enhanced spectra were observed irrespective of the polarization state of the incident light, in contrast to the enhancement by the excitation of SPP. The results thus obtained indicate that the enhancement stems from the excitation of the collective electron resonance. Moskovits suggested in an early work that the enhancement of Raman signals from molecules adsorbed on roughened electrodes was the direct result of the excitation of the collective electron resonance in submicroscopic bumps on the

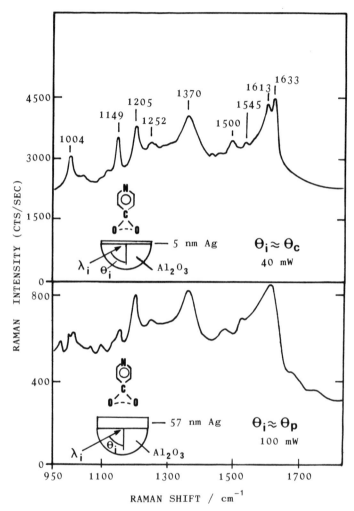

Figure 2. Surface enhanced Raman spectra of isonicotinic acid chemisorbed on a 5 nm Ag island film (top) and chemisorbed on a 57 nm Ag film (bottom). Laser light at 514.5 nm was incident at the critical angle for total internal reflection (top) and at the resonance angle for the SPP excitation (bottom) [Reprinted from R. Dornhaus, R. E. Benner, R. K. Chang, and I. Chabay, *Surf. Sci.* **101**, 367 (1980)].

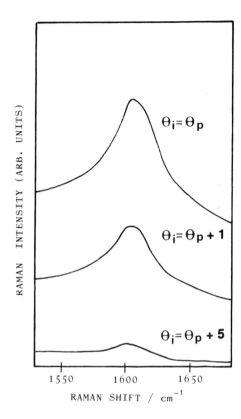

Figure 3. Incident angle dependence of the intensity of the 1600 cm^{-1} Raman line from 4-nitrobenzoic acid chemisorbed on a 57 nm Ag film. The intensity of the Raman line is enhanced by at least one order of magnitude with the excitation at the angle of SPP resonance, i = 514.5 nm [Reprinted from R. Dornhaus, R. E. Benner, R. K. Chang, and I. Chabay, *Surf. Sci.* **101**, 367 (1980)].

surface of electrochemically processed electrodes.[15] Island films and colloids of Ag and Au exhibit a broad absorption band in the visible and near-infrared regions arising from the excitation of the collective electron resonance. Chen *et al.*[34] investigated the enhancement of Raman signals by the presence of Ag and Au island films and observed the wavelength dependence of Raman intensity from species on the Ag island film. They found that the wavelength dependence was consistent with the broad absorption peak at ~750 nm of the Ag island film, which is due to the excitation of the collective electron resonance. The collective electron resonance thus takes an important part in the enhancement of Raman signals.

Besides the collective electron resonance, the plasma resonance by isolated metal particles gives rise to a strong EM field at rough metal surfaces. Several groups have treated theoretically the enhancement of the local field using a model either of a sphere or spheroid metal particle on a flat surface or of a hemisphere protruding from a conductive plane.[35–39] Takemori *et al.*,[39] for example, approximated the roughened metal electrode surface with a sphere placed above a flat metal surface. Their calculation showed that a strong field could be generated as a result of the coupling of the sphere plasmon with the SPP of the substrate. For the incidence of *p*-polarized light, Raman signals from a linear molecule on a Ag sphere with a 10 nm radius, which is placed 11 nm above the Ag substrate, may be enhanced by a factor of 10^8. The interaction between metal particles and metal protrusions, which gives rise to collective electron resonances, is neglected in this type of treatment. The neglected interaction can be included in the averaged local field calculation.

Bergman and Nitzan[40] treated molecules adsorbed on island films as Raman scatterers embedded in an isotropic composite material and calculated the averaged local field intensities experienced by the Raman scatterers in the Maxwell–Garnett and in the effective medium theories for the purpose of estimating the enhancement factor. The calculation shows that the averaged enhancement ratio may be as large as two orders of magnitude when the volume fraction of Ag is 0.1. The averaged internal field in two-component composites was also discussed by Aspnes.[41] Girourad *et al.* observed the transmittance of Cu/MgF_2 films in the wavelength region 0.3–0.9 μm and compared it with the theoretical calculations in the Maxwell–Garnett and Bruggeman theories.[42] The observed results are in agreement with both of the theoretical estimations as long as the volume fraction of metal is small. The Bruggeman theory gives more satisfactory results than the Maxwell–Garnett theory for large volume fractions of metal.

The EM enhancement is long-range, while the "chemical" mechanism is operating only on molecules in direct contact with the metal surface. The distance dependence of the enhancement factor can provide us with crucial information to distinguishing the EM mechanism from the "chemical" one. Kerker *et al.*[37] and Murray *et al.*[43] investigated the distance dependence. Kovacs *et al.*[44] studied the distance dependence of the enhancement factor using a lightly substituted phthalocyanine as the sample molecule and Langmuir–Blodgett films of arachidic acid as the spacer layer between the sample molecules and In or Ag metal island films. The magnitude of the enhancement could also be determined in this measurement, because a few monolayers of phthalocyanine on substrates without metal islands gave fairly good normal Raman spectra. The results obtained experimentally are in good quantitative agreement with the calculations based on EM theory used by Aroca and Martin[45] for both the distance dependence and the magnitude of enhancement. This shows that Raman enhancement in this system is attributable to the EM mechanism.

Cavities in a metal film are mirror images of metal spheres and spheroids on the metal film, and may give rise to the enhancement of the local EM field. Cold-deposited Ag films have highly porous structure and exhibit very strong SERS signals.[5,46,47] Chew and Kerker treated the cavity as a small sphere or oblate located near the surface in the metal phase.[48] Their calculation showed that an enormous total enhancement as large as 10^{11} may be obtained for oblate spheroidal cavities. The role of cavity sites such as conical wedges and enclosures between small particles in SERS was discussed by Gersten and Nitzan.[49] They showed that the latter are associated with a particularly large enhancement of local field intensity. Later, Takemori and Inoue showed that a strong local field could be excited on the part of a spherical cavity surface close to the metal (Ag) surface. A SERS enhancement factor of 10^4 was calculated for a spherical cavity of 10 nm radius and allocated to a portion of the inner surface of the cavity near the outer surface of the metal layer.[50]

The enhancement of Raman signals from pyridine adsorbed on Ag films deposited on a cold substrate in UHV was studied experimentally by Seki et al.[51,52] They concluded that the cavities and small pores in the silver films were the location of active site of SERS, because the enhancement effect of the cavities was reduced drastically by filling them with Xe atoms.[52] Osawa et al.[53] computed the averaged electric field intensity within island and porous metal films evaporated on the basal planes of ATR prisms of LaSF15 glass in the effective medium theory, and the results were compared with the observed intensity of a Raman band of copper phthalocyanine (CuPC) deposited on the silver films. Figure 4 shows the results obtained with Ag films of various thicknesses. The calculated intensities of X- and Y-components of the electric field within the composites normalized to the intensity of the incident light (E_0^2), that is, the enhancement factor of the field, are shown versus the angle of incidence in Fig. 4a–c. The Z-component, normal to the metal surface, is not shown in these drawings because it is very weak. The normalized electric field intensity of SPP, perpendicular to the surface, is shown together with the observed Raman intensities in Fig. 4d.

The observed angular dependence of Raman intensity is in good qualitative agreement with that of the electric field intensity in the composite layer. This means that most of the intensity of the observed Raman signal is attributable to molecules present in the composite layer—in other words, the molecules present in the pores and crevices of the Ag films, as well as those adsorbed on the very irregular outer surfaces of the films, contribute most of the observed Raman intensity. Quantitatively speaking, however, there is a severe discrepancy between the calculated field strength and the observed Raman intensity. A large enhancement beyond 8×10^2 is observed in Raman intensity

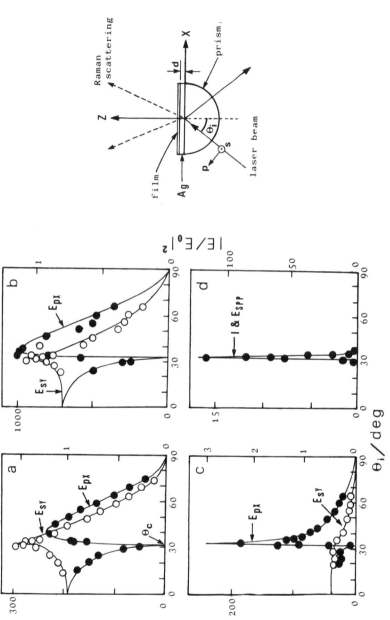

Figure 4. Left, intensity of the 1530 cm^{-1} Raman line from a 2 nm thickness CuPC film as a function of the angle of incidence. Thickness of Ag underlayers is (a) 5, (b) 10, (c) 30, and (d) 50 nm. Open and closed circles represent Raman intensity measured with s- and p-polarized laser light, respectively. The solid curves in (a)-(c) show the calculated strength of electric field within the composite layers (see text). The solid curve in (d) represents the calculated incident angle dependence of the SPP field strength. Right, the Kretschmann ATR configuration employed in the SERS measurement [Reprinted from M. Osawa and W. Suétaka, *Surf. Sci.* **186**, 583 (1987)].

of CuPC deposited on a 20 nm Ag film, while the intensity of the averaged electric field in the composite layer is comparable to that of the incident light.

The discrepancy may be attributed to the increase in the scattering volume, the molecular orientation, the chemical enhancement, or the enhancement of the local electric field at the cavity sites. The thin films used in the measurements were porous and the real surface area of the films should be much larger than the apparent geometrical surface area. However, the observed enormous enhancement of Raman intensity cannot be explained by the increase in surface area alone: Eickmans et al. estimated that the roughness factor of a cold deposited Ag film of 200 nm in thickness reached a value of only 20 at 120 K.[54] Similarly small enhancement due to molecular orientation effects is also expected. The "chemical" enhancement may give rise to the observed intensity increase. However, CuPC is believed to be immune to the "chemical" enhancement in the measured system.[29] The observed enhancement of Raman signals can, therefore, be ascribed to the strong local EM field at crevices and enclosures among bumps.

Colloidal silver or gold particles show absorption of visible light due to the excitation of collective electron resonance, resulting in the emergence of the characteristic colors of the sols. The EM enhancement of Raman signals can be expected for species adsorbed on the colloidal particles through the excitation of collective electron resonance. Creighton et al. investigated the SER spectra of pyridine adsorbed on colloidal silver and gold particles.[55] The absorption band of the Ag sol has two peaks, and the low frequency component, which is assignable to aggregates of very small colloidal particles, increases in intensity and shifts to longer wavelengths with an increase in the dimension of aggregates. They found that the intensity of Raman signals increased when a laser beam of a wavelength in the low frequency absorption was incident.

Kerker et al.[56] measured Raman spectra of citrate ions adsorbed on colloidal silver and found that the enhancement increased monotonically with increasing wavelength of the incident light until 647.1 nm. Kerker and Wang discussed the enhancement of Raman signals of species adsorbed on colloidal metal particles in the framework of the electrodynamic mechanism.[57,58] The model calculation shows that the EM field near a small metal particle is strongly enhanced at frequencies where the plasma resonance is excited within the metal particles. The measured enhancement of Raman signals, thus, results from the superimposition of the increase in strength of the EM field, which induces the Raman scattering, on the intensity increase arising from the excitation of a similar resonance by the Raman shifted light. They also discussed the possibility of an enormous enhancement of the signals of coherent anti-Stokes

Raman scattering of molecules located near the surface of a small silver particle.[59]

2.2. "Chemical" Mechanisms

Although the EM mechanism evidently takes an important part in the enhancement of Raman signals from surface species, there are several observations that cannot be explained with the EM mechanism alone. Very high enhancement factors have been calculated for particular cases of the EM enhancement in some theoretical works. However, important interactions between metal particles or protrusions and damping of the field that would substantially reduce the calculated enhancement are often not taken into account in these calculations. The enhancement factor generally accepted at present for the EM mechanism is 4 orders of magnitude at its maximum. The measured enhancement factor extends to 10^6 or more. The electronic configuration of chemical species changes more or less upon adsorption, leading to the change in the Raman cross section. Surface enhanced Raman spectra of species in direct contact with the surface are often different from those of species in the condensed phase in relative intensity and in position of Raman bands. There is additional evidence for the enhancement operating on species bound directly with the surface. For example, SER spectra of species adsorbed on electrodes often show a steep increase in the intensity of Raman signals of adsorbed species at a specified electrode potential, and the potential of the maximum Raman intensity changes depending upon the wavelength of the incident laser beam. These observations, which cannot be explained with the EM mechanism, strongly suggest the existence of an additional mechanism in the signal enhancement of SERS. This additional mechanism is generally termed the "chemical" mechanism and has been reviewed comprehensively by Otto.[2,8,9] It is generally accepted as an important mechanism of the enhancement, although the detailed mechanism is still in debate.

The observations mentioned above suggest that the additional mechanism may be the resonance Raman effect (Section 2.1 of Chapter 5). However, most of the species exhibiting SER spectra are not absorbing in the visible region. The formation of a surface complex upon adsorption on the metal substrate and the resonance Raman scattering in the complex therefore, has been proposed by several groups.[60-65] Furtak and Miragliotta observed enhanced Raman scattering from pyridine-silver complexes deposited on a Pt electrode.[65] The observed spectra were consistent with that of pyridine adsorbed at Lewis acid sites on the Ag substrate, indicating that the enhancement could be attributed to the resonant Raman effect in the pyridine–Ag^+ complex. This model of the surface complex formation, however, may be included in the framework of adsorption-assisted resonance Raman scattering.

The "chemical" mechanism involves the charge transfer between the adsorbates and the metal substrates and can be divided into ground-state and excited-state mechanisms.[66] In the latter model or in the metal electron-mediated resonance Raman effect,[9] a charge-transfer transition from metal states below the Fermi level to unoccupied orbitals (called affinity levels) of adsorbates, shifted and broadened upon adsorption, is assumed.

Persson presented a resonance Raman scattering model involving the chemisorption-induced charge transfer excitation.[67] Similar models have been proposed by Adrian,[68] Billmann and Otto,[69] Burstein et al.[70] and Ueba et al.[71] Ueba et al. drew the energy diagram of pyridine adsorbed on Ag and explained the SERS as the adsorbate-induced resonant Raman scattering via charge transfer from filled metal states to the first affinity level of adsorbed pyridine. In the resonance enhancement mechanism, the enhancement of Raman signals thus takes place when the electronic excitations of charge transfer become resonant with the incident light. It follows that, when an adsorbate-metal system gives SERS, it should exhibit an absorption band in the visible region (1.5–3.0 eV). Demuth and Sanda[72] assigned a new excitation centered at 1.9 eV above the Fermi level of Ag appearing in the electron energy loss spectrum of pyridine chemisorbed on a Ag substrate to the adsorption-induced charge transfer between adsorbed pyridine and silver. They observed the charge-transfer bands in the visible region for a number of molecules adsorbed on single crystals and cold deposited Ag surfaces.[73,74] On the other hand, only few optical observations of the charge-transfer excitations have been reported. Yamada et al.[75] measured absorption spectra of β- and γ-picolines and pyridine adsorbed on Ag island films and found new weak absorption bands around 600 nm. They assigned the weak bands to the charge-transfer excitations. The information on the affinity levels can also be obtained in the measurement of inverse photoemission.[9]

In the ground-state charge oscillation mechanism, a charge transfer between the ground state of the adsorbate and the metal states is assumed.[76–78] The charge transfer is modulated by molecular vibrations and thus brings about Raman bands. The enhancement of the Raman cross section comes from intensity-borrowing from the metal, which has a high scattering cross section. Consequently, the enhancement factor of this mechanism would be rather modest. It has also been deduced that atomic-scale roughness is required in most cases for the emergence of the enhancement in this mechanism, and that the enhancement arising from this mechanism should show no dependence on the excitation wavelength except for certain dye molecules.[66] However, this mechanism is less prevalent than the other and is not treated further.

The strong enhancement of Raman signals has been observed for species adsorbed on electrochemically roughened electrodes and cold deposited metal surfaces, while adsorbed molecules on clean and smooth low-index faces of Ag single crystals show unenhanced or weakly enhanced Raman spectra. Furthermore, enhanced Raman signals from species adsorbed on electrode surfaces often decrease irreversibly in intensity when the electrode potential is shifted. These facts suggest the existence of active sites associated with the "chemically" enhanced Raman scattering. Furtak *et al.* attributed the enhancement of the Raman intensity of pyridine and SCN⁻ ions adsorbed on Ag electrodes to complexes of these species formed at defects of the electrode surfaces.[63] In a later work, an electron-deficient Ag site was deduced as the active site.[65]

Otto *et al.* proposed the "ad-atom hypothesis," assuming the charge-transfer via the tunneling of the electron or hole through the ad-atom from the metal into electronic states of the adsorbate.[79,80] Later, they replaced the hypothesis with the concepts of "adsorption sites of lowered affinity level" and "sites of increased surface-electron-photon coupling," termed E(extra)-site and N(normal)-site, respectively.[8,9] Their position on the active sites is briefly outlined in the following.

The N-sites are present on smooth metal surfaces such as atomically smooth low-index faces of Ag single crystals and well-annealed evaporated Ag film surfaces, and molecules are physisorbed on this type of site. The adsorbate is not stuck on a particular location of the surface, and the enhancement stems from non-local charge transfer excitation. The affinity levels of the N-species (species adsorbed on the N-sites) is higher than that of the E-species. When the affinity level of physisorbed molecules is close to the resonance, the resonance Raman effect takes place, but the frequencies of the skeletal vibrations of the N-species are nearly unshifted from those of free molecules.

The E-sites are present on rough metal surfaces such as cold deposited Ag films, Ag island films, and electrochemically processed electrodes. The adsorption on the E-sites results in the shift of vibrational bands of the adsorbed or chemisorbed species. Oxygen adsorbs preferentially at the E-sites, passivating the SERS activity of silver island films and cold deposited silver films as a result.[81]

In contrast to the EM mechanism, the enhancement of Raman signals by the "chemical" mechanism is confined to the species in direct contact with the metal surface. Since the "chemical" enhancement is a kind of resonance Raman effect, it has chemical specificity, the relative intensity of Raman bands may change in the SER spectra, and the enhancement factor changes significantly from system to system. In addition, the "chemical" enhancement may take place for species adsorbed on solid substrates other than free-electron metals. The adsorption on transition metals, however, gives rise to high affinity levels and the resulting enhancement is low. The enhanced Raman spectra of species adsorbed on ZnO, TiO_2,[82] and PbTe[83] were reported. It should be added here that there are several works in which enhanced Raman scattering has been predicted on the basis of the EM mechanism for molecules embedded in dielectric particles,[84] and the enhancement of Raman signals of CuPC on the surfaces of GaP particles was also attributed to the EM mechanism.[85]

3. Applications

The enhancement of Raman signals by the EM mechanism has been found for species on Ag, Au, Cu, In, Na, and Al. The observation of SERS has also appeared in the literature for species on Cd, Hg, Ni, Pd, and Pt.[44] Furthermore, the deposition of a small amount of SERS-inactive metal on EM-SERS-active metal has been used to investigate the adsorption on the SERS-inactive metal. On the other hand, deposition of a silver island film is used to enhance the Raman signals of surface layers of solid materials. The observation of SER spectra is thus applied to various fields of fundamental and applied sciences.

3.1. Electrode Surfaces and Corrosion

As mentioned earlier, SERS is useful for investigating electrode surfaces and metallic corrosion, because SER signals from water are generally very weak and the interference of solution species is insignificant even when a conventional electrochemical cell is used. There are excellent review articles of the application of SERS to the investigation of electrode surfaces.[86,87]

A large number of organic and inorganic compounds are employed as corrosion inhibitors for minimizing the corrosion of metals. Vibrational spectroscopy should be a powerful tool for investigating the mechanism of corrosion inhibition and for finding new effective inhibitors. Infrared reflection measurements therefore, have been employed for the identification of chemical species in the protective films formed on metal surfaces in the reaction with the inhibitors and for the determination of film thickness. Inhibitors reduce the corrosion rate of metals by the adsorption at chemically active sites on the metal surfaces in addition to the formation of protective films. Since most of the infrared measurements have been performed *ex situ,* as described in Section 1.4 of Chapter 2, direct information about adsorption-mediated corrosion inhibition has scarcely been obtained. Surface enhanced Raman scattering should be more useful than infrared spectroscopy in the investigation of adsorption-mediated corrosion inhibition, because it has high sensitivity and is easily applicable to the *in situ* observation of the behavior of inhibitors on metal surfaces in corrosive media.

Benzotriazole (BTA) is widely used as an effective corrosion inhibitor of copper and copper-based alloys and has been a subject of SERS observations.[88–92] Surface enhanced Raman spectra of BTA and its derivatives adsorbed on copper were measured *in situ* in aqueous solutions containing H_2SO_4 and Na_2SO_4. Spectra of adsorbed BTA were obtained at potentials lower than –0.2 V (versus SCE), and the formation of its complex with Cu was found at potentials higher than –0.3 V by Youda *et al.*[91] They concluded that the adsorption-mediated and protective film-controlling corrosion inhibition occurs at potentials lower and higher than ~0.25 V, respectively. They also observed SER spectra of pyridine adsorbed on iron deposited on an electrochemically roughened Ag electrode. Good parallelism was seen at varying electrode potential between the Raman intensity of pyridine adsorbed on iron and the inhibition of iron corrosion by this compound.

The relation between the molecular structure of organic compounds and their inhibitive action against the corrosion of copper was studied by means of the SERS observation and electrochemical measurements using BTA and structurally related compounds.[89] Marked differences were found in the ability of the compounds to inhibit the corrosion, and these were correlated with the molecular structure of the compounds. The compounds investigated were adsorbed molecularly except for the loss of protons. Mercaptobenzimidazole and hydroxybenzimidazole showed the maximum and minimum abilities to inhibit the corrosion, respectively, and the difference in ability was attributed to the number and type of functional groups and atoms capable of interacting with copper. The poor inhibitive ability of the latter was also ascribed to the tautomerism to form the keto form.[89]

Fleischmann *et al.*[90] investigated the efficiency of inhibitors (BTA and its related compounds) in the corrosion of Cu in aqueous chloride media. A relation was found between the rate of loss of Raman intensity of adsorbed inhibitors at open circuit and the rate of corrosion of Cu. In other words, stable Raman intensity at open circuit implies effective corrosion inhibition. This implies that the corrosion of copper would be reduced by the stable adsorption of inhibitors at chemically active sites on the metal surface. The results obtained also show that localized corrosion can be monitored through the simultaneous observation of the Raman bands of the inhibitor and aggressive anions bound to the surface. The relative strength of adsorption bonding of the inhibitors, which corresponds to the inhibitive ability of the compounds as mentioned above, was estimated by using appropriate mixtures of inhibitors and observing SER spectra of adsorbed species.

Mercaptobenzothiazole (MBT) is another effective corrosion inhibitor for copper and several other metals and alloys. Mercaptobenzothiazole in aqueous solutions take the form of ionized thiol or thione (Fig. 5), depending upon the pH of the solution: the latter is predominant at pH values lower than 4 and the former at pH > 9. In the intermediate region of 4 < pH > 9, both of the species are present in the solutions. Unexpectedly, SER spectra of MBT adsorbed on a silver electrode were invariant independent of the pH of the solution and were in good agreement with the normal Raman spectrum of MBT in a basic solution, as can be seen in Fig. 6. The spectrum of MBT in the thione form, spectrum B in Fig. 7, is remarkably different from spectra in Fig. 6B and C. Figure 6 thus indicates that MBT is adsorbed on the Ag electrode in the ionized-thiol form irrespective of the pH of the solution.

Halide ions give rise to a marked decrease in the corrosion inhibitive action of MBT. The SER spectrum of MBT adsorbed on a Ag electrode changes strikingly upon addition of KI, as can be seen in Fig. 7, indicating the conversion of the ionized-thiol species into the thione species.[93] The observation of SER spectra thus provides direct information on the effect of halide ion addition to the adsorbed inhibitor molecules. The loss of inhibitive action in the thione form is significant, because the poor inhibitive ability of hydroxybenzimidazole is ascribed to the presence of keto form.[89]

Molybdate and tungstate ions are efficient inhibitors of local corrosion of metals, even in aggressive media containing chloride ions. The pH of the solution is decreased in the pits and crevices of metal surfaces, and oligomers

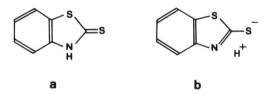

a   **b**

Figure 5. (a) Thione and (b) ionized thiol forms of mercaptobenzothiazole.

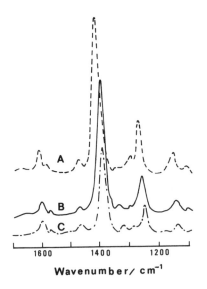

Figure 6. Raman spectra of MBT in a basic solution (A, pH = 10.0), and adsorbed on Ag electrodes in neutral (B, pH = 6.0) and in acidic (C, pH = 2.0) solutions. Concentration of MBT: (A): 0.1 mol/1, (B) and (C): 1×10^{-7} mol/l [Reprinted from M. Ohsawa and W. Suëtaka, *J. Electron Spectrosc. Relat. Phenom.* **30**, 221 (1983)].

are formed from the above-mentioned oxoacid ions when the pH of the solution is lowered. Patterson and Allen[94] observed SER spectra of paramolybdate and paratungstate A, the first oligomers formed from molybdate and tungstate monomers, respectively, chemisorbed irreversibly on copper in order to investigate the behavior of the oxoacids in surface defects, which can be considered the early stage of local corrosion. The obtained spectra showed that the adsorbed oligomers interact with the copper surface via terminal oxygens and

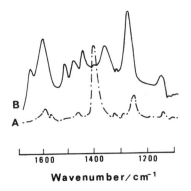

Figure 7. Surface enhanced Raman spectra of MBT adsorbed on a Ag electrode in an acidic (pH = 2.0) Na_2SO_4 solution (A) before and (B) after (C) the addition of KI. Concentration of MBT, 1×10^{-7} mol/l. Electrode potential, −0.2 V (versus SCE) [Reprinted from M. Ohsawa and W. Suëtaka, *J. Electron Spectrosc. Relat. Phenom.* **30**, 221 (1983)].

heal flaws formed in the passive film, thus avoiding the growth of the localized corrosion. The reduction of the oligomers on the copper electrodes was also monitored by the SERS measurement, and the results obtained indicated that the reduced species also adsorb on copper and play a role in the inhibition of the localized corrosion.

Thiocyanate ion has properties similar to halide ions and causes the breakdown of the passivity of iron. The change of surface oxide film upon addition of SCN^- ion was monitored *in situ* with Raman spectroscopy, and the replacement of oxide and hydroxide with iron thiocyanate, which is non-passivating, was observed.[95]

Surface enhanced Raman spectra of oxide films on metal electrodes were observed after the electrodeposition of a silver overlayer.[96] Melendres *et al.* observed *in situ* SER spectra of passive films on nickel,[97] iron, and chromium[98] with the same technique. Whereas $Fe(OH)_2$ and Fe_3O_4 were present in the prepassive region of iron electrodes, $FeOOH$ was found in the passive region. On chromium electrodes, Cr_2O_3 and $Cr(OH)_3$ were detected in aqueous chloride media. The conversion of these compounds to a higher oxide was observed in the transpassive region. $Ni(OH)_2$ and NiO were found in the passive films on nickel electrodes. The former was formed in the prepassive region and the latter upon further anodic sweep of the electrode potential.

The reaction of oxide-covered metal surfaces with the environment is the initial process of corrosion.[99] Dorain *et al.* investigated the protonation process of the oxide on Ag electrodes.[100–102] When the electrode potential was swept cathodically, adsorbed O^{2-} was reduced to adsorbed OH^- and, in some cases, further to form H_2O. When chromate ion, having a strong corrosion inhibition ability through oxide layer formation, was present in the solution, adsorbed OH^- was not protonated to form H_2O, and the adsorbed OH^- was deprotonated to form an oxide layer upon addition of O_2 gas to the solution.[102] In contrast, in the presence of MnO_4^-, which causes an increase in the corrosion rate of Ag, the oxide layer was protonated to form OH^- and, at very low pH, to form adsorbed H_2O upon a cathodic sweep of the electrode potential.[100] The behavior of the chromate ions observed in this work is probably relevant to the mechanism of the chromate-passivation of metal surfaces.[102] The oxidation and reduction of Ag electrodes in alkaline solutions is also an interesting subject in connection with batteries. Iwasaki *et al.*[103] observed SER spectra of species on Ag electrodes in a NaOH aqueous solution. The measured spectra showed the presence of adsorbed oxygen and hydroxy species. The former, which was unstable and present only in a narrow potential region, was reduced to adsorbed OH species in the cathodic sweep of the electrode potential. At very negative potentials, Raman features from adsorbed water were observed.

Photoelectrochemical oxidation of Ag_2O to AgO was also observed *in situ* by SERS.

Surface enhanced Raman spectra of phosphate adsorbed on Ag electrodes have revealed that the fractions of adsorbed PO_4^{3-}, HPO_4^{2-}, and $H_2PO_4^-$ change depending upon the electrode potential and consequently upon the pH at the electrode/electrolyte interface. The effective pH values at the interface therefore, can be estimated from the SER spectra of adsorbed phosphates.[101]

Ethylenediaminetetraacetic acid (EDTA) forms chelate complexes with a variety of metal ions and plays an important role in electroless plating. Bunding Lee *et al.* investigated the behavior of EDTA on copper and silver electrodes.[104] When Cu was used as the electrode, EDTA exhibited no SER spectrum, although the electrode surface was SERS active. This implies that Cu-EDTA complexes are highly soluble and the concentration of EDTA is very low on Cu electrodes. Surface enhanced Raman spectra of EDTA on the Ag electrode changed significantly with the change in electrolyte. In nitrate and dilute chloride solutions, spectra of adsorbed EDTA resembled the solution spectrum of of EDTA. However, in a 0.1 M KCl solution, SER spectra of adsorbed species showed the decarboxylation of EDTA at the electrode surface.

Surface enhanced Raman spectroscopy is a potential tool for investigating electrochemical reactions at electrode surfaces. However, the intense surface EM field can be supported only by a discrete set of metal surfaces that have been appropriately roughened. Leung and Weaver electrodeposited 1–3 monolayers of Pt and Pd onto gold in an effort to overcome this problem and were successful in observing enhanced Raman spectra of CO adsorbed on the transition metals.[105] Gold was chosen as the SERS-supporting substrate in this work, because it could be used in wide potential ranges and was suitable for deposition of a variety of metals and oxides. Although the electronic structures of the deposited metals are not the same as in the bulk, valuable information would be obtained from the observation of SER spectra of species on the deposited metals. The same workers extended this technique to the observation of adsorption on Rh, Ru, and metal oxide films.[106,107] This technique is particularly advantageous in the observation of low frequency regions, in comparison with infrared spectroscopy. In the observation of SER spectra of adsorbed CO on the Rh and Ru films, the M–C stretching bands of adsorbed CO and the M–O stretching bands of surface oxides were located.[106] Figure 8 shows representative SER spectra for rhodium-coated gold in 0.1 M $HClO_4$ saturated with CO. A band appeared at about 460 cm^{-1} in the positive-going potential excursion at initial and moderately positive potentials, but disappeared at potentials more positive than 0.7 V, where voltammetric electrooxidation of CO was observed. This band was ascribed to the Rh–CO stretching

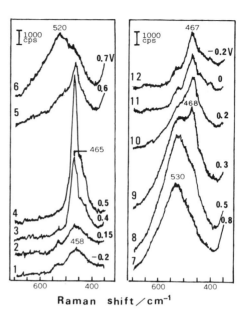

Figure 8. Potential-dependent SER spectra in the low-frequency (300–700 cm^{-1}) region for a Rh-coated Au electrode in 0.1 M HClO$_4$. Spectra were recorded at the indicated potentials (versus SCE) sequentially as numbered. Laser excitation was 70 mW at 647.1 nm [Reprinted with permission from L.-W. H. Leung and M. J. Weaver, *Langmuir* **4**, 1076 (1988). © 1988 American Chemical Society].

vibration. A broad band appeared at about 520 cm^{-1} at positive potentials in the positive-going potential excursion and decreased in intensity in the negative-going excursion. This band was also present in the absence of CO and was assigned to the surface Rh–O stretching vibration.

In certain electrochemical systems, deposition of a small amount of metal takes place at potentials more positive than the equilibrium electrode potential for the metal couple. This process is called underpotential deposition (UPD).[108,109] Underpotential deposition is often used to form a submonolayer or a few monolayers of a metal on foreign metal electrodes. Leung *et al.* compared SER spectra of several species adsorbed on underpotentially deposited Ag and Cu on roughened Au electrodes with spectra of the same species on unmodified Au, Ag, and Cu electrodes. The intensity of the Raman signals of species adsorbed on the UPD electrodes was comparable to or higher than those of the corresponding spectra of the bulk metal electrodes. Several Raman bands of species on UPD silver were shifted to higher frequencies than those on bulk silver electrodes. This shift was attributed to the change in the electron polarization of the adsorbate on the UPD silver.[110] The SER spectra of adsorbed species at Fe deposited on Ag was investigated by Aramaki *et al.* as already mentioned.[91,111] In SER spectra of roughened silver surfaces, low frequency signals were observed and attributed to Ag-Ag vibrations.[112,113] Li *et al.* measured SER spectra from UPD silver on a gold substrate in chloride aqueous media and observed the surface Au-Au and Au-Ag stretching bands in addition to signals from surface chloride complexes.[114]

When the electronic structure of the surface layer of a SERS-active metal changes appreciably through the deposition of foreign metals, the surface EM field on the metal may be reduced in intensity. In addition, the adsorption at SERS-active sites on the surfaces may be blocked by the UPD metal, thus resulting in the quenching of SERS signals. This effect was studied by several groups,[115–117] and the damping of SERS signals from adsorbates was found as a result of UPD of Cd, Tl, and Pb on silver substrates. For example, Kester observed the complete quenching of Raman signals from BTA before the completion of underpotential deposition of Tl on Ag electrodes and attributed it to the displacement of BTA from the Ag surface.[117] Guy *et al.*[118] observed that the SERS signal from pyridine adsorbed on a Ag electrode in chloride aqueous media was substantially reduced in intensity by a 70% coverage with a Pb monolayer. In a later work, they measured Raman intensity-Pb coverage profiles using the same system and compared the measured profile with the theoretical one calculated using a model of EM enhancement at overlayer-covered metal ellipsoids. They concluded that the enhancement of the observed system should be ascribed to a "chemical" mechanism, because the observed and calculated profiles were in poor agreement.[119]

Since the substrate surface is irradiated by a strong laser beam during the measurement of SERS, the irradiated surface may undergo photoalteration. The intensity of Raman signals from thiocyanate and chloride ions adsorbed on silver electrodes increased by the illumination of the electrode during the oxidation-reduction cycles (ORC) for the preparation of SERS-active electrodes.[120] This was also the case for copper electrodes in aqueous halide solutions, and a tenfold increase in Raman intensity was observed as a result of illumination during ORC. Copper halides (CuX, X = halogen) are changed into metallic copper and CuX_2 in the presence of light. CuX_2 may be dissolved in the solution and the metallic Cu formed may act as a nucleation center for the electrochemical reduction of CuX to copper. Scanning electron micrograph observation revealed that the morphology of the electrode surface was changed by the irradiation during ORC. The SERS intensity increase was attributed, therefore, to the change in surface morphology or the formation of SERS-active sites on the irradiated electrode surface.[121] The surface morphological effect on SERS intensities was also reported by others.[122,123] The magnitude of the irradiation effect is also dependent on the wavelength of the radiation during ORC.[120] Tuschel and Pemberton[124] made similar measurements, yet deconvoluting the effect of ORC irradiation from that of irradiation for Raman excitation, and confirmed that the increase in SERS intensity was dependent on the frequency of the laser light used.

Surface enhanced Raman spectroscopy was utilized for the investigation of photochemical reactions on semiconductor electrodes. The incident laser light played dual parts in the excitation of Raman scattering and in the induction of the photochemical reaction. Since p-type InP is a promising material for photocells, the redox reactions of methylviologen (MV) on a p-InP electrode was studied with electrochemical techniques and SERS measurements.[125] The SERS measurement was carried out using a p-InP electrode covered with a Ag island film. The photovoltaic response of the electrode remained unchanged by the deposition of the Ag islands. The photoreduction of MV^{2+} proceeding at the electrode was monitored *in situ* with Raman scattering, and the signals of adsorbed reduction products $MV^{\cdot+}$ and MV^0 were obtained. Figure 9 shows

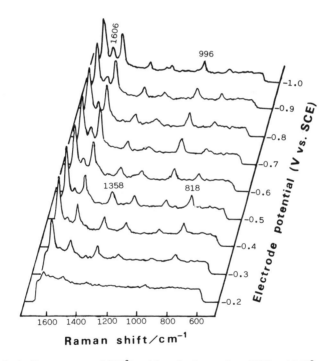

Figure 9. *In situ* Raman spectra of MV^{2+} and its reduction products $MV^{\cdot+}$ and MV^0 on a silver modified p-InP electrode as a function of electrode potential. The spectra were obtained by using the 363.8 nm line of an argon ion laser [Reprinted with permission from Q. Feng and T. M. Cotton, *J. Chem. Phys.* **90**, 983 (1986). © 1986 American Chemical Society].

surface enhanced resonance Raman (SERR) spectra from a Ag-modified p-InP electrode at different electrode potentials. Raman bands of MV^{2+} are very weak and no band of this species can be seen in the spectrum recorded at -0.2 V (versus SCE). Raman bands of $MV^{\cdot+}$ appear at 818 and 1358 cm^{-1} at a potential of -0.3 V where MV^{2+} in solution begins undergoing reduction to $MV^{\cdot+}$ at the InP electrode. New Raman bands appear at 996 and 1606 cm^{-1} with a further decrease in the electrode potential, accompanying the intensity loss of the $MV^{\cdot+}$ bands. The new bands are assignable to adsorbed MV^0. Only scarcely discernible Raman bands of $MV^{\cdot+}$ were observed by using a bare p-InP electrode under the same experimental conditions, although the cyclic voltammogram showed both $MV^{2+}/MV^{\cdot+}$ and $MV^{\cdot+}/MV^0$ peaks.

Time-resolved SER spectra have been measured using an optical multichannel analyzer in combination with the cyclic voltammetry or the potential step method.[101,126] For instance, short-lived intermediates formed on electrodes during electrochemical reactions were detected in time-resolved SER spectra.[127] Gao et al.[128] obtained real-time SER spectra from species on gold electrodes subjected to cyclic voltammetry using a Kr$^+$ laser and a SPEX Model 1877 spectrometer mounted with a charge-coupled devices (CCD) detector. The molecular-level information on the adsorbates can be obtained at various stages of the cyclic voltammetry with this technique. They investigated electrooxidation of sulfide on Au electrodes in aqueous alkaline and acidic media.

Potential modulation methods have been employed in order to detect weak Raman signals of surface species and faint changes in intensity and peak frequency of Raman bands caused by the potential step.[129,130] Tian et al.[131] used high frequency (100–1 kHz) potential modulation combined with a rather long integration time (~0.5 s) for recording spectra in their potential modulation SERS measurement. In this method, averaged spectra are recorded instead of difference spectra as is usual in potential modulation methods. They applied this technique to observe SER spectra of SCN$^-$ ions adsorbed on Ag electrodes. The use of a high modulation frequency reduces the removal of surface active sites at negative electrode potentials, resulting in the increase in Raman intensity at negative potentials.

Surface oxide-free aluminum may support the EM field of SPP in the visible and near-ultraviolet regions. Abraham et al.[132] prepared Ag/Al composite electrodes through the evaporation of a very thin and continuous Ag layer in UHV onto an Al film on a quartz substrate. The composite electrode showed a reflectivity minimum of p-polarized light due to the SPP excitation. The electrodes exhibited the reflectivity minimum even after the repeated anodic polarizations, although the minimum was shifted to shorter wave vectors. Care must be taken to avoid the peeling off of the metal film from the substrate when the composite electrode is immersed in the solution. This is the case whenever a metal film evaporated on a semiconductor or dielectric substrate is used as the electrode. One way to prevent this is to use a four-electrode arrangement, in which an auxiliary electrode made of the same metal as the working one is used. The potential of the auxiliary electrode is set in the range where no reaction occurs at the electrode. The working thin film electrode is then connected to the auxiliary one and immersed in the solution.[132] Holze[133] proposed a new and simple electrochemical method for the preparation of a gold electrode for SERS observation. This method minimizes the trapping of molecules or ions present near the electrode during its activation procedure and allows observation of spectra free from spectroscopic artifacts.

Since Raman scattering is excited with visible light and the scattered light also has wavelengths in the visible region, optical fibers can be employed for monitoring remote environments through Raman spectroscopy.[134,135] However, Raman lines of the fiber material and fluorescence from the media surrounding the fiber may give rise to serious problems. This problem may be overcome by taking advantage of the surface enhancement of Raman signals. Mullen and Carron[136] obtained good SER spectra of a probe molecule (cobalt phthalocyanine) using abrasively roughened optical fibers onto which Ag was deposited. This method will be feasible for water pollutant analysis. Cyr *et al.*[137] applied fiber optic Raman spectroscopy to the *in situ* observation of SER spectra of pyridine adsorbed in a microscope area of a copper electrode surface utilizing a specially designed electrochemical cell shown in Fig. 10.

3.2. Thin Films and Adsorbed Species

Surface enhanced Raman spectroscopy has also been employed in the investigation of solid surfaces in air, in reactant gas, and in vacuum. However, the information obtained from the measurement of SER spectra changes depending upon the enhancement mechanism. When Raman signals are enhanced by the "chemical" mechanism, the enhanced signals come from the species directly bound to the metal surface. When the EM mechanism is operating in the enhancement of surface Raman spectra, the enhanced EM field should extend beyond the species directly bound to the metal.

The enhanced surface field due to the SPP excitation decays exponentially over a few hundred nanometers into the vacuum. The field of plasma resonance by metal particles and of relating collective resonance may decay more rapidly than that of SPP with distance from the surface of island films. Theoretical calculations show that the enhanced EM field on a Ag sphere decays sharply with distance from the surface, provided the diameter of the sphere is much

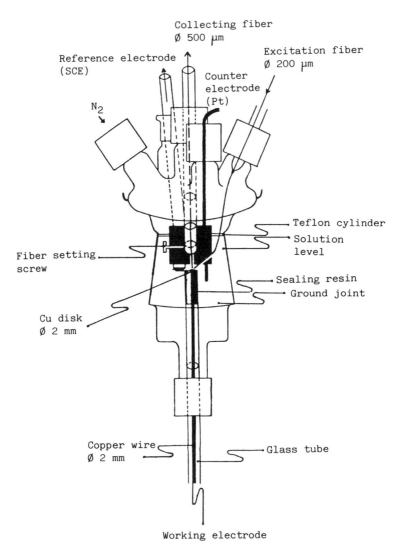

Figure 10. Schematic diagram of the cell used in the fiber optic measurement of SERS [Reprinted from A. Cyn, P. Plaza, M. Jouan, E. Laviron, and Nguyen Quy Dao, *C. R. Acad. Sci. Ser. II* **314**, 43 (1992), Editions Gauthier-Villars].

smaller than the wavelength of the exciting light.[37] The enhancement dependent upon the distance from the surface of a hemispheroid protruding from a conducting plane was also estimated.[38] The often-quoted distance dependence of the enhanced field strength for a molecule on a metal sphere is $\sim(a/r)^{12}$, where a is the radius of the sphere and r is the molecule-sphere center distance.[138] When the sphere is coated with a monolayer, the Raman intensity will fall off as $\sim(a/r)^{10}$.[139] These relations are also applicable to rough metal surfaces, if a is taken as the local radius of curvature of the hemispherical protruding and r as the molecule-curvature center distance.

Allara et al.[140] showed that when continuous Ag films, which could be modelled as round silver blobs on a silver substrate, were deposited on polymer layers, the enhanced Raman signals of the polymer came from molecules at distances as far as 20 nm from the Ag/polymer interface. On the other hand, when polymer films were spin cast on the top of films of Ag islands, which had very sharp edges, on glass substrates, the saturation of Raman intensity occurred below 2 nm polymer thickness. Similar characteristics were observed in the SER spectra of Langmuir–Blodgett (LB) films.[141] The SER spectra of the LB films of cadmium arachidate deposited on grating metal surfaces, where an enhanced field of SPP was generated, changed markedly with evaporation of an island overlayer of Ag, suggesting the irregularity in the outer layer of the LB film. These observations are in agreement with the theoretical prediction and indicate that the strong EM field of SPP generally extends to a greater distance from the metal surface than that generated on metal island films.

Information about the outermost layers is thus obtained through the deposition of a Ag island film onto the sample surface, provided the layer remains unchanged by the Ag deposition. The deposition of the Ag island film therefore, is widely used for the observation of Raman spectra of surface layers of polymers and inorganic materials. The analysis of surface layers of polystyrene spheres[142] and latex[143] was carried out by the observation of SER spectra of the silver-island coated samples. Attention must be paid, however, because the deposited Ag islands may be contaminated by impurities such as carbon and oxygen. The contaminants of vapor-deposited Ag films were characterized by Del Priore et al.[144] They concluded from XPS and Raman measurements that the Ag films deposited at a pressure of 3×10^{-4} Pa were covered by carbon and oxygen species at a coverage of the order of several monolayers, and the contaminants were composed principally of graphite and silver oxide. Nanba and Martin observed SER spectra of Ag and Cu smoke particles and detected an oxide layer composed of M_2O and MO as well as graphitic carbon.[145] The use of Au island films instead of Ag islands, in combination with an incident laser light of long wavelength, may be advisable for avoiding the interference from the Raman bands of oxide contaminants.

Adsorption of NO_2 gas at LB monolayers of dysprosium and holmium bisphthalocyanine complexes on gold island films were investigated with surface enhanced resonance Raman spectroscopy.[146] The SERR spectra of the neat LB layers and the adducts implied the formation of a charge-transfer complex upon the adsorption and the coordination of NO_2 to the metal center in the phthalocyanine complex.

Watanabe et al.[147] measured Raman spectra of carbon deposited on Si[100] surfaces by plasma glow discharge. The obtained spectra exhibited the optical phonon modes of a wurtzite SiC layer whose thickness was about 2 nm. They deduced the enhancement of Raman signals by a factor of 170 and claimed to have discovered a new type of enhanced Raman scattering.

Molecular level information on the mechanism of lubrication has been obtained through the measurement of infrared reflection and emission spectra, as described in the preceding chapters but SERS also provides useful information on the mechanism. Organic sulfides and disulfides are used as effective extreme pressure agents. Sandroff et al.[148] observed Raman spectra of diphenyl disulfide and diphenyl sulfide adsorbed on silver island films and postulated the formation of mercaptides on the metal surfaces. However, the results obtained in this work are not in agreement with the mechanism generally accepted by tribologists who postulate the formation of mechanically weak metal compounds at the contact.[149] The discrepancy may come from the difference between the reaction conditions of the extreme pressure agents with metals: under practical conditions, the agent is applied for reducing the friction under high speed and high load condition, and the applied agent reacts with the metal at high temperatures.

The observation of SER spectra has been used for the investigation of various species at gas/metal interfaces. Typical species investigated include pyridine, hydrocarbons, fluorinated hydrocarbons, CO, and CO_2.[5,8] Sandroff et al.[150] observed SER spectra of the radical cation of tetrathiafulvalene (TTF) in a non-electrochemical system. Tetrathiafulvalene is oxidized upon adsorption onto silver and gold island films to form the radical cation $TTF^{\cdot+}$. The oxidation of TTF, a typical electron donor, is accompanied by a charge transfer to the adsorbent metal, resulting in the shift of the TTF vibrational bands. The same workers proposed the use of SER spectra of TTF and related compounds adsorbed on SERS-active metals for the investigation of the charge transfer from adsorbates to metals because of a rather wide range of combination of ionization potentials and work functions which can be employed.

The change in vibrational frequency of CO adsorbed on metals resulting from the coadsorption of various species has been extensively studied by infrared reflection spectroscopy. Yamamoto and Nanba[151] observed a shift in the Raman band due to the C–O stretching vibration of CO adsorbed on Ag. The coadsorption of Xe and Kr induced shifts to lower frequencies and that of oxygen and chlorine to higher ones. The obtained results are in good agreement with the infrared observations. The Raman spectra of CO on copper films having atomic-scale surface roughness were observed by Akemann and Otto.[152] They observed four vibrational modes of a C–O stretching, a restricted (frustrated) rotation, and restricted translations normal (Cu–CO stretching) and parallel to the local surface, and assigned these bands to CO molecules adsorbed at surface defect sites.

Jiang and Campion[153] evaporated silver adatoms onto a Rh[100] face in ultrahigh vacuum and observed SER spectra from pyridine chemisorbed on the Ag adatoms. They estimated an enhancement factor of ~40 ± 25 and attributed most of the enhancement to the "chemical" mechanism. Wolkow and Moskovits[154] observed the electron energy loss spectrum and the SER spectrum of benzene adsorbed on cold deposited silver films. The results obtained with the EELS measurement did not reconcile with the SERS observations. Whereas EELS implied the existence of two types of adsorption sites on the silver films, SERS results did not strongly imply the existence of such adsorption sites. They have deduced, therefore, that Raman signals of strongly bound benzene species are much more strongly enhanced than those of weakly bound species and the difference in selection rules operating in EELS and SERS also gives rise to the change in the observed spectra.

Krasser et al.[155] observed SER spectra from benzene adsorbed on Ni/SiO_2 and Pt/SiO_2 clusters. The obtained spectra exhibited some Raman-inactive modes in addition to the strongly enhanced Raman-active modes of adsorbed benzene. Although the Raman-active modes were only slightly shifted in peak frequency from liquid values, the relative intensities changed drastically and the change was dependent on the wavelength of the incident laser beam. In addition, metal-carbon vibrations appeared in the obtained spectra, showing a strong interaction between the adsorbate and the metal. Coinage metals, which support a strong surface EM field, were not used in the measurement. The obtained results, therefore, imply that the "chemical" mechanism takes a dominant part in the enhancement of the Raman scattering from the observed system.

The presence of a static electric field enhances the efficiency of additional Raman scattering from the longitudinal optical (LO) phonon in semiconductors arising from the long-range interaction of the phonon with the electronic system. This enhanced scattering, called the electric field-induced Raman scattering (EFIRS), stems, in III-V compounds, mainly from a tilting of the energy bands as a result of surface band bending.[155] The same workers observed EFIRS from the LO phonon of GaAs covered with a thin Sb layer and investigated the electronic band bending in the interface region of the semiconductor resulting from coverage of foreign materials. Since the structure of the overlayer affects the band bending, they also observed Raman scattering from the Sb overlayer and discussed the band bending dynamics.

Surface selection rules, which constitute the basis for the determination of the molecular orientation of surface species, are somewhat complicated for SERS, because rough surfaces are generally used in the observation of SER spectra and Raman signals are enhanced by multiple mechanisms. The selection rules for SERS have been reviewed by Creighton[156] and Otto et al.,[9] but a broad consensus on the rules has not yet been reached. In a first approximation, the surface selection rules can be formulated on the basis of the EM mechanism. However, the observed change in relative intensity of Raman bands often cannot be explained by the EM selection rules.[9] Furthermore, Raman-inactive modes often appear in measured SER spectra. The appearance of these modes has been explained by the reduction of molecular symmetry or the formation of surface complexes. Moskovits et al.[157] observed SER spectra of aromatic molecules adsorbed on vacuum-deposited silver films. Several Raman-inactive modes appeared in the measured spectra and were explained by the steep change in the radiative electric field near sharp features on the surface. In this mechanism, the modes belonging to the same representation as the components of the dipole-quadrupole tensor become Raman-active.[157] They also observed a change in the spectrum following the illumination by the laser beam and attributed this to molecular reorientation on the heated surface. Surface enhanced Raman spectra of C_{60} and C_{70} fullerenes vapor-deposited on rough Ag films were measured by Akers et al.[158] Besides Raman-active lines, several Raman-inactive lines appeared in the measured spectra, due to either the reduction in molecular symmetry as a result of the interaction with the surface or the rapid spatial rate of change in the radiative field in the vicinity of the rough metal surface.

Pemberton et al.[159] formulated a simple method for the determination of the orientation of low-symmetry adsorbates on metal surfaces on the basis of the EM enhancement mechanism. The validity of this method was proven by the confirmation of the established orientation of methanol adsorbed on metal surfaces. Figure 11a shows a Raman spectrum of methanol in the bulk liquid and SER spectrum of methanol adsorbed on a roughened Ag electrode. The molecular orientation determined from the measured spectra is shown in Fig.

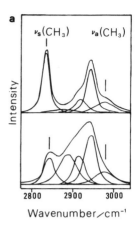

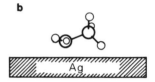

Figure 11. (a) Raman spectra in the C–H stretching vibration region of methanol containing 0.4 M LiBr in bulk liquid (top) and at a roughened Ag electrode surface (bottom); (b) proposed methanol orientation at the electrode [Reprinted with permission from J. E. Pemberton, M. A. Bryant, R. L. Sobocinski, and S. L. Joa, *J. Phys. Chem.* **96**, 3776 (1992), © 1992 American Chemical Society].

11b. This approach was then applied to the determination of the orientation of several alcohols and thiols adsorbed on Ag and Au surfaces.

Surface enhanced Raman spectra of p-aminobenzoic acid (PABA) on silver-coated alumina were measured in air and in several solvents.[160] In air and in nonpolar solvents, the PABA molecule was adsorbed with its ring parallel to the surface of the substrate. On the other hand, in polar solvents, PABA was adsorbed through the COO⁻ group with its ring obliquely oriented to the substrate surface because of the hydrogen bond formation between the amino group of PABA and the solvent molecules.

The near-infrared light from a Nd:YAG laser has been used to excite SERS of species on Cu and Au as well as Ag, because the strong EM field can be excited on these metals in the near-infrared region as well, and "fluorescence-free" SER spectra can be obtained. Jennings *et al.* observed SER spectra of species on gold island films using an FT-Raman spectrometer combined with the near-infrared excitation.[24] The obtained SER spectra of vanadylphthalocyanine and 3,4,9,10-perylenetetracarboxylic dianhydride were in agreement with previously obtained data using visible excitation. An enhancement factor of 100 was estimated and attributed to the EM enhancement. The obtained results thus indicate that FT-SERS will be a powerful tool for observing vibrational spectra of fluorescent materials.

The use of silver needle substrates may facilitate the observation of SER spectra in practical applications. However, the fabrication of the needle substrates for the efficient excitation of SPP is difficult, because a uniform arrangement of sharp needles is necessary for the excitation of the maximum strength field. Wachter *et al.*[161] proposed a new method for fabricating the optimum silver needle substrates. They deposited a monolayer of microspheres on a glass microscope slide and then a suitable thickness of silver was evaporated at a grazing angle onto the prepared microsphere base layer. Surface enhanced Raman spectra of benzoic acid on the needle substrates fabricated with this method are shown in Fig. 12. Silver needles were grown on naturally rough $InSnO_2$, on a monolayer of 39-nm diameter latex microspheres and on 75-nm latex microspheres. Laser beams polarized parallel (Z) and perpendicular (E) to the major axis of the needle were used for the excitation of Raman scattering.

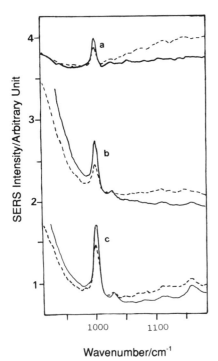

Figure 12. Surface enhanced Raman spectra of benzoic acid on Ag needles formed on (a) $InSnO_2$, (b) 39-nm latex microsphere and (c) 75-nm microsphere. The solid line trace was obtained using Z-mode excitation and the dashed trace E-mode excitation [Reprinted with permission from E. A. Wachter, A. K. Moore, and J. W. Haas, III, *Vibrational Spectrosc.* **3**, 73 (1992)].

The strongest Raman signals were recorded with the 75-nm latex microsphere substrate.

The backscattering Kretschmann-ATR configuration is often used for observing SER spectra of species on SPP-supporting metal films. However, the efficiency of the generally used collection optics of Raman scattered light is low, because the hemicylindrical prism used in the ATR configurations does not allow the efficient collection of Stokes-shifted light emitted to the prism side. For the efficient collection of the scattered light, Wittke *et al.*[162] proposed the use of a Weierstrass prism, which allows the collection of all of the light emitted to the prism side, in place of the generally used hemicylindrical prism. By comparison with the external backscattering configuration, an enhancement by a factor of two orders of magnitude was attained in the observation of Raman signals from carbon contamination on silver films.

References

1. T. E. Furtak and R. K. Chang, eds., *Surface Enhanced Raman Scattering,* Plenum, New York (1982).
2. A. Otto, in: *Light Scattering in Solids,* Vol. 4 (M. Cardona and G. Güntherodt, eds.), Springer, Berlin (1984), p. 289.
3. H. Metiu and P. Das, *Ann. Rev. Phys. Chem.* **35,** 507 (1984).
4. H. Metiu, *Prog. Surf. Sci.* **17,** 153 (1984).
5. I. Pockrand, *Surface Enhanced Raman Vibrational Studies at Solid/Gas Interfaces,* Springer, Berlin (1984).
6. M. Moskovits, *Rev. Mod. Phys.* **57,** 783 (1985).
7. A. Campion, in: *Vibrational Spectroscopy of Molecules on Surfaces* (J. T. Yates, Jr., and T. E. Madey, eds.), Plenum, New York (1987), p. 345.
8. A. Otto, *J. Raman Spectrosc.* **22,** 743 (1991).
9. A. Otto, I. Mrozek, H. Grabhorn, and W. Akemann, *J. Phys. Condens. Matter* **4,** 1143 (1992).
10. M. Fleischmann, P. J. Hendra, and A. J. McQuillan, *Chem. Phys. Lett.* **26,** 163 (1974); A. M. McQuillan, P. J. Hendra, and M. Fleischmann, *J. Electroanal. Chem.* **65,** 933 (1975).
11. D. L. Jeanmaire and R. P. Van Duyne, *J. Electroanal. Chem.* **84,** 1 (1977).
12. M. G. Albrecht and J. A. Creighton, *J. Am. Chem. Soc.* **99,** 5215 (1977).
13. M. R. Philpott, *J. Chem. Phys.* **62,** 1812 (1975).
14. R. P. Van Duyne, *J. Phys.* (Paris) **38,** C5-239 (1977).
15. M. Moskovits, *J. Chem. Phys.* **69,** 4159 (1978); *Solid State Comm.* **32,** 59 (1979).
16. M. K. Debe, *Prog. Surf. Sci.* **24,** 1 (1987).
17. C. Nylander, B. Liedberg, and T. Lind, *Sens. Actuators* **3,** 79 (1982/83).
18. S. Löfås and B. Johnsson, *Chem. Commn.* **21,** 1526 (1990).
19. E. M. Yeatman and E. A. Ash, *Electron Lett.* **23,** 1091 (1987).
20. W. Hickel and W. Knoll, *Thin Solid Films* **187,** 347 (1990).
21. T. Okamoto and I. Yamaguchi, *Opt. Comm.* **93,** 265 (1992).
22. S. M. Angel and D. D. Archibald, *Appl. Spectrosc.* **43,** 1097 (1989).

23. M. Fleischmann, D. Sockalingum, and M. M. Musiani, *Spectrochim. Acta* **A46,** 285 (1990).
24. C. A. Jennings, G. J. Kovacs, and R. Aroca, *J. Phys. Chem.* **96,** 1340 (1992).
25. W. Knoll, M. R. Philpott, J. D. Swalen, and A. Girlando, *J. Chem. Phys.* **77,** 2254 (1982).
26. J. C. Tsang, J. R. Kirtley, and J. A. Bradley, *Phys. Rev. Lett.* **43,** 772 (1979).
27. A. Girlando, M. R. Philpott, D. Heitmann, J. D. Swalen, and R. Santo, *J. Chem. Phys.* **72,** 5187 (1980).
28. S. Hayashi, *Surf. Sc.* **158,** 229 (1985).
29. R. Koh, S. Hayashi, and K. Yamamoto, *Solid State Comm.* **64,** 375 (1987).
30. Y. J. Chen, W. P. Chen, and E. Burstein, *Phys. Rev. Lett.* **36,** 1207 (1976); W. P. Chen, G. Ritchie, and E. Burstein, *Phys. Rev. Lett.* **37,** 993 (1976).
31. B. Pettinger, A. Tadjeddine, and D. H. Kolb, *Chem. Phys. Lett.* **66,** 544 (1979).
32. R. Dornhaus, R. E. Benner, R. K. Chang, and I. Chabay, *Surf. Sci.* **101,** 367 (1980).
33. S. Ushioda and Y. Sasaki, *Phys. Rev.* **B27,** 1401 (1983).
34. C. Y. Chen, I. Davoli, G. Ritchie, and E. Burstein, *Surf. Sci.* **101,** 363 (1980).
35. P. K. Aravind and H. Metiu, *J. Phys. Chem.* **86,** 5076 (1982).
36. P. K. Aravind, R. Rendell, and H. Metiu, *Chem. Phys. Lett.* **85,** 396 (1982).
37. M. Kerker, D-S. Wang, and H. Chew, *Appl. Opt.* **19,** 4159 (1980).
38. J. Gersten and A. Nitzan, *J. Chem. Phys.* **73,** 3023 (1980); *Surf. Sci.* **158,** 165 (1985).
39. T. Takemori, M. Inoue, and K. Ohtaka, *J. Phys. Soc. Jpn.* **56,** 1587 (1987).
40. D. J. Bergman and A. Nitzan, *Chem. Phys. Lett.* **88,** 409 (1982).
41. D. E. Aspnes, *Phys. Rev. Lett.* **48,** 1629 (1982).
42. F. E. Girourad, V-V. Truong, and T. Yamaguchi, *Appl. Surf. Sci.* **33/34,** 959 (1988).
43. C. A. Murray, D. L. Allara, and M. Rhinewine, *Phys. Rev. Lett.* **46,** 57 (1981).; C. A. Murray and D. L. Allara, *J. Chem. Phys.* **76,** 1290 (1982).
44. G. J. Kovacs, R. O. Loutfy, P. S. Vincett, C. Jennings, and R. Aroca, *Langmuir* **2,** 689 (1986).
45. R. Aroca and F. Martin, *J. Raman Spectrosc.* **16,** 156 (1985).
46. H. Seki and T. J. Chuang, *Chem. Phys. Lett.* **100,** 393 (1983).
47. E. V. Albano, S. Daiser, G. Ertl, R. Miranda, K. Wandelt, and N. Garcia, *Phys. Rev. Lett.* **51,** 2314 (1983).
48. H. Chew and M. Kerker, *J. Opt. Soc. Am.* **132,** 1025 (1985).
49. J. I. Gersten and A. Nitzan, *Surf. Sci.* **158,** 165 (1985).
50. T. Takemori and M. Inoue, *J. Phys. Soc. Jpn.* **57,** 3188 (1988).
51. H. Seki, T. J. Chuang, J. F. Escobar, H. Morawitz, and E. V. Albano, *Surf. Sci.* **158,** 254 (1985).
52. H. Seki, T. J. Chuang, M. R. Philpott, E. V. Albano, and K. Wandelt, *Phys. Rev.* **B31,** 5533 (1985).
53. M. Osawa and W. Suëtaka, *Surf. Sci.* **186,** 583 (1987); M. Osawa, S. Yamamoto and W. Suëtaka, *Appl Surf. Sci.* **33/34,** 890 (1988).
54. J. Eickmans, A. Otto, and A. Goldmann, *Surf. Sci.* **171,** 415 (1986).
55. J. A. Creighton, C. G. Blatchford, and M. G. Albrecht, *J. Chem. Soc. Faraday II* **75,** 790 (1979).
56. M. Kerker, O. Siiman, L. A. Bumm, and D-S. Wang, *Appl. Opt.* **19,** 3253 (1980).
57. D.-S. Wang and M. Kerker, *Phys. Rev.* **B24,** 1777 (1981).
58. M. Kerker, *Acc. Chem. Res.* **17,** 271 (1984).
59. H. Chew, D.-S. Wang, and M. Kerker, *J. Opt. Soc. Am.* **B1,** 56 (1984).
60. B. Pettinger, U. Wenning, and D. M. Kolb, *Ber. Bunsenges. Physik. Chem.* **82,** 1326 (1978).
61. A. Regis and J. Corset, *Chem. Phys. Lett.* **70,** 305 (1980).
62. T. Watanabe, O. Kawanami, K. Honda, and B. Pettinger, *Chem. Phys. Lett.* **102,** 565 (1983).

63. T. E. Furtak and D. Roy, *Phys. Rev. Lett.* **50,** 1301 (1983).
64. J. E. Pemberton, *J. Electroanal. Chem.* **167,** 317 (1984).
65. T. E. Furtak and J. Miragliotta, *Surf. Sci.* **167,** 381 (1986).
66. M. E. Lippitsch, *Phys. Rev. Lett.* **B29,** 3101 (1984).
67. B. N. J. Persson, *Chem. Phys. Lett.* **82,** 561 (1981).
68. F. J. Adrian, *J. Chem. Phys.* **77,** 5302 (1982).
69. J. Billmann and A. Otto, *Solid State Comm.* **44,** 105 (1982).
70. E. Burstein, Y. J. Chen, C. Y. Chen, S. Lundquist, and E. Tosatti, *Solid State Comm.* **29,** 567 (1979).
71. H. Ueba, S. Ichimura, and H. Yamada, *Surf. Sci.* **119,** 433 (1982).
72. J. E. Demuth and P. N. Sanda, *Phys. Rev. Lett.* **47,** 57 (1981).
73. Ph. Avouris and J. E. Demuth, *J. Chem. Phys.* **75,** 4783 (1981).
74. D. Schmeisser, J. E. Demuth, and Ph. Avouris, *Chem. Phys. Lett.* **87,** 324 (1982).
75. H. Yamada, K. Toba, and Y. Nakao, *J. Electron Spectrosc. Relat. Phenom.* **45,** 113 (1987); H. Yamada, H. Nagata, K. Toba, and Y. Nakao, *Surf. Sci.* **182,** 269 (1987).
76. F. R. Aussenegg and M. E. Lippitsch, *Chem. Phys. Lett.* **59,** 114 (1978).
77. S. L. McCall and P. M. Platzmann, *Phys. Rev.* **B22,** 1660 (1980).
78. H. Abe, K. Manzel, W. Schulz, M. Moskovits, and D. P. DiLella, *J. Chem. Phys.* **74,** 792 (1981).
79. J. Billmann, G. Kovacs, and A. Otto, *Surf. Sci.* **92,** 153 (1980).
80. A. Otto, *Appl. Surf. Sci.* **6,** 309 (1980).
81. I. Mrozek and A. Otto, *J. Electron Spectrosc. Relat. Phenom.* **54/55,** 895 (1990).
82. H. Yamada and Y. Yamamoto, *Surf. Sci.* **134,** 71 (1983).
83. J. E. Potts, R. Merlin, and D. L. Partin, *Phys. Rev.* **B27,** 3905 (1983).
84. D-S. Wang, M. Kerker, and H. Chew, *Appl. Opt.* **19,** 2315 (1980).
85. S. Hayashi, R. Koh, Y. Ichiyama, and K. Yamamoto, *Phys. Rev. Lett.* **60,** 1085 (1988).
86. R. L. Birke and J. R. Lombardi, in: *Spectroelectrochemistry: Theory and Practice* (R. J. Gale, ed.), Plenum, New York (1988).
87. D. A. Weitz, M. Moskovits, and J. A. Creighton, in: *Chemistry and Structure at Interfaces—New Laser and Optical Techniques* (R. B. Hall and A. B. Ellis, eds.), VCH, Deerfield Beach, FL (1986), p. 197.
88. J. T. Kester, T. E. Furtak, and A. J. Bevolo, *J. Electrochem. Soc.* **129,** 1716 (1982).
89. D. Thierry and C. Leygraf, *J. Electrochem. Soc.* **132,** 1009 (1985).
90. M. Fleischmann, I. R. Hill, G. Mengoli, M. M. Musiani, and J. Akhavan, *Electrochim. Acta* **30,** 879 (1985).
91. R. Youda, N. Nishihara, and K. Aramaki, *Corrosion Sci.* **28,** 87 (1988).
92. K. T. Carron, G. Yue, and M. L. Lewis, *Langmuir* **7,** 2 (1991).
93. M. Ohsawa and W. Suëtaka, *J. Electron Spectrosc. Relat. Phenom.* **30,** 221 (1993).
94. M. L. Patterson and C. S. Allen, *Appl. Surf. Sci.* **18,** 377 (1984).
95. C. A. Melendres, J. Acho, and R. L. Knight, *J. Electrochem. Soc.* **138,** 877 (1991).
96. J. C. Rubin and J. Dunnwald, *J. Electroanal. Chem.* **258,** 327 (1989).
97. C. A. Melendres and M. Pankuch, *J. Electronal. Chem.,* in press.
98. C. A. Melendres, M. Pankuch, Y. S. Li, and R. Knight, *Electrochim. Acta,* in press.
99. For example: J. E. Crowell, J. G. Chen, D. M. Hercules, and J. T. Yates, Jr., *J. Chem. Phys.* **86,** 5804 (1987).
100. P. B. Dorain, *J. Phys. Chem.* **90,** 5808, 5812 (1986)
101. P. B. Dorain, K. U. von Raben, and R. K. Chang, *Surf. Sci.* **148,** 439 (1984).
102. P. B. Dorain and J. L. Bates, *Langmuir* **4,** 1269 (1988).

103. N. Iwasaki, Y. Sasaki, and Y. Nishina, *J. Electrochem. Soc.* **135**, 2531 (1988); *Surf. Sci.* **198**, 524 (1988).
104. K. A. Bunding Lee, G. McLennan, and J. G. Gordon, *Ber. Bensenges. Physik. Chem.* **91**, 305 (1987).
105. L-W. H. Leung and M. J. Weaver, *J. Am. Chem. Soc.* **109**, 5113 (1987).
106. L-W. H. Leung and M. J. Weaver, *Langmuir* **4**, 1076 (1988).
107. D. Gosztola and M. J. Weaver, *Langmuir* **5**, 776 (1989).
108. D. M. Kolb, in: *Advances in Electrochemistry and Electrochemical Engineering*, Vol. 11 (P. Delahay and C. W. Tobias, eds.), Wiley, New York (1978), p. 125.
109. R. A. Adzic, in: *Advances in Electrochemistry and Electrochemical Engineering*, Vol. 13, Wiley, New York (1984), p. 159.
110. L-W. H. Leung, D. Gosztola, and M. J. Weaver, *Langmuir* **3**, 45 (1987).
111. K. Aramaki and J. Uehara, *J. Electrochem. Soc.* **137**, 185 (1990); J. Uehara, H. Nishijima, and K. Aramaki, *J. Electrochem. Soc.* **137**, 2677 (1990).
112. I. Pockrand; *Chem. Phys. Lett.* **85**, 37 (1982).
113. D. Roy and T. E. Furtak, *Chem. Phys. Lett.* **124**, 299 (1986).
114. J. Li, J. Liang, C. Korzeniewski, and S. Pons, *Langmuir* **3**, 470 (1987).
115. B. Pettinger and L. Moerl, *J. Electron Spectrosc. Relat. Phenom.* **29**, 383 (1983).
116. T. Watanabe, K. Yanagihara, K. Honda, B. Pettinger, and L. Moerl, *Chem. Phys. Lett.* **96**, 649 (1983).
117. J. J. Kester, *J. Chem. Phys.* **78**, 7466 (1983).
118. A. L. Guy, B. Bergami, and J. E. Pemberton, *Surf. Sci.* **150**, 226 (1985).
119. A. L. Guy and J. E. Pemberton, *Langmuir* **3**, 777 (1987).
120. F. Barz, J. G. Gordon II, M. R. Philpott, and M. J. Weaver, *Chem. Phys. Lett.* **91**, 291 (1982).
121. D. Thierry and C. Leygraf, *Surf. Sci.* **149**, 592 (1985).
122. D. V. Murphy, K. U. von Raben, T. T. Chen, J. F. Owen, R. K. Chang, and B. L. Laube, *Surf. Sci.* **124**, 529 (1983).
123. T. M. Devine, T. E. Furtak, and S. H. Macomber, *J. Electroanal. Chem.* **164**, 299 (1984).
124. D. D. Tuschel and J. E. Pemberton, *Langmuir* **4**, 58 (1988).
125. Q. Feng and T. M. Cotton, *J. Chem. Phys.* **90**, 983 (1986).
126. M. J. Weaver, F. Barz, J. G. Gordon II, and M. R. Philpott, *Surf. Sci.* **125**, 409 (1983).
127. C. Shi, W. Zhang, R. L. Birke, and J. R. Lombardi, *J. Phys. Chem.* **94**, 4766 (1990).
128. X. Gao, Y. Zhang, and M. J. Weaver, *Langmuir* **8**, 668 (1992).
129. W. Suëtaka and M. Ohsawa, *Appl. Surf. Sci.* **3**, 118 (1979).
130. M. Fleischmann, P. R. Graves, and J. Robinson, *J. Electroanal. Chem.* **182**, 73 (1985).
131. Z. Q. Tian, W. F. Lin, and B. W. Mao, *J. Electroanal. Chem.* **319**, 403 (1991).
132. M. Abraham, J. P. Rolland, A. Tadjeddine, and G. Schiffmacher, *Surf. Sci.* **146**, 351 (1984).
133. R. Holze, *Surf. Sci.* **202**, L612 (1988).
134. S. D. Schwab and R. L. McCreery, *Appl. Spectrosc.* **41**, 126 (1987).
135. J. M. Bello and T. Vo-Dinh, *Appl. Spectrosc.* **44**, 63 (1990).
136. K. I. Mullen and K. T. Carron, *Anal. Chem.* **63**, 2196 (1991).
137. A. Cyr, P. Plaza M. Jouan, E. Laviron, and Nguyen Quy Dao, *C. R. Acad. Sci. Ser. II* **314**, 43 (1992).
138. S. L. McCall, P. M. Platzman, and O. A. Wolff, *Phys. Lett.* **77A**, 381 (1980).
139. T. M. Cotton, R. H. Uphaus, and D. J. Möbius, *J. Phys. Chem.* **90**, 6071 (1986).
140. D. L. Allara, C. A. Murray, and S. Bodoff, in: *Physico-chemical Aspects of Polymer Surfaces* Vol. 1 (K. L. Mittal, ed.), Plenum, New York (1983), p. 33.
141. W. Knoll, M. R. Philpott, and W. G. Golden, *J. Chem. Phys.* **77**, 219 (1982).

142. P. Barnickel and A. Wokaum, *Mol. Phys.* **67,** 1355 (1989).
143. M. C. Buncick, R. J. Warmack, and T. E. Ferrell, *J. Opt. Soc. Am.* **34,** 927 (1987).
144. L. V. Del Priore, C. Doyle, and J. D. Andrade *Appl. Spectrosc.* **36,** 69 (1982).
145. T. Nanba and T. P. Martin, *Phys. Stat. Sol.* **(a)76,** 235 (1983).
146. J. Souto, R. Aroca, and J. A. De Saja, *J. Raman Spectrosc.* **22,** 787 (1991).
147. M. Watanabe, H. Minagawa, T. Miyazaki, and T. Yamashina, *Surf. Sci.* **208,** 164 (1989).
148. C. J. Sandroff and D. R. Hershchbach, *J. Phys. Chem.* **86,** 3277 (1982).
149. For example: F. P. Bowden and D. Taber, *The Friction and Lubrication of Solids,* Clarendon, Oxford (1954).
150. C. J. Sandroff, D. A. Weitz, J. C. Chung, and D. R. Herschbach, *J. Phys. Chem.* **87,** 2127 (1983).
151. I. Yamamoto and T. Nanba, *Surf. Sci.* **202,** 377 (1988).
152. W. Akemann and A. Otto, *J. Raman Spectrosc.* **22,** 797 (1991).
153. X. Jiang and A. Campion, *Chem. Phys. Lett.* **140,** 95 (1987).
154. R. A. Wolkow and M. Moskovits, *J. Chem. Phys.* **96,** 3966 (1992).
155. W. Krasser, J. Kojnok, and J. Geurts, *J. Raman Spectrosc.* **22,** 777 (1991).
156. J. A. Creighton, in: *Spectroscopy of Surfaces,* Vol. 16 (R. J. H. Clark and R. E. Hester, eds.), Wiley, Chichester (1988), p. 37.
157. M. Moskovits, D. P. DiLella, and K. J. Maynard, *Langmuir* **4,** 67 (1988).
158. K. L. Akers, L. M. Cousins, and M. Moskovits, *Chem. Phys. Lett.* **190,** 614 (1992).
159. J. E. Pemberton, M. A. Bryant, R. L. Sobocinski, and S. L. Joa, *J. Phys. Chem.* **96,** 3776 (1992).
160. J. B. Bello, V. Anantha Narayanan, and T. Vo-Dinh, *Spectrochim. Acta* **48A,** 563 (1992).
161. E. A. Wachter, A. K. Moore, and J. W. Haas, III, *Vibrational Spectrosc.* **3,** 73 (1992).
162. W. Wittke, A. Hatta, and A. Otto, *Appl. Phys.* **A48,** 289 (1989).

Index